LIFE COMES FROM SPACE
The Decisive Evidence

LIFE COMES FROM SPACE
The Decisive Evidence

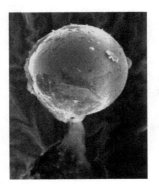

Milton Wainwright
University of Sheffield, UK
N Chandra Wickramasinghe
University of Buckingham, UK

Foreword by
Gensuke Tokoro

World Scientific

NEW JERSEY · LONDON · SINGAPORE · BEIJING · SHANGHAI · HONG KONG · TAIPEI · CHENNAI · TOKYO

Published by

World Scientific Publishing Co. Pte. Ltd.

5 Toh Tuck Link, Singapore 596224

USA office: 27 Warren Street, Suite 401-402, Hackensack, NJ 07601

UK office: 57 Shelton Street, Covent Garden, London WC2H 9HE

British Library Cataloguing-in-Publication Data
A catalogue record for this book is available from the British Library.

LIFE COMES FROM SPACE
The Decisive Evidence

ISBN 978-981-126-625-6 (hardcover)
ISBN 978-981-126-626-3 (ebook for institutions)
ISBN 978-981-126-627-0 (ebook for individuals)

For any available supplementary material, please visit
https://www.worldscientific.com/worldscibooks/10.1142/13140#t=suppl

Biological Entities from Space.

The thought that life originated on this insignificantly little and comparatively unimportant sphere to me seems so inconceivably egotistical. After the Earth cooled of the great heat of its assemblage, life units came to it through space, into which they had been thrown from some other more developed sphere or spheres, Reaching the earth, they adapted themselves to the environment they found here; and then began the evolution of the various species as we have them each growing individual being a collection of cell communes... It is not impossible that, when we find the ultimate unit of life, we shall learn that the journey through far space never could harm it and that there is very little that could stop it... Such units of life have come, and possibly still are coming, without injury through the cold of space.

Thomas Alva Edison

Foreword

Milton Wainwright's stratospheric sampling has led to the recovery of what he calls "biological entities" which are microscopic structures whose sizes, shapes as well as elemental composition strongly suggest their biological nature. The body of evidence discussed in this book gives credence to a space origin of all such entities, and thus to the continuing link between the terrestrial biosphere and a vast external cosmic biosphere.

These findings are backed up by amazing images of fossilized biology in the North West African and Polonnaruwa meteorites which have been published in peer reviewed journals. This evidence has been ignored, simply because it would mean the abandonment of a culturally constrained belief that life (or at any rate our type of life) is confined to our planet Earth. Such a belief is part of a pre-Copernican philosophy that is being fiercely defended half a millennium after scientists had finally accepted that the Earth is not the centre of the Universe. The time has now come to concede that life is not centred on our planet but is a cosmic phenomenon, as has been meticulously documented by Chandra Wickramasinghe and Fred Hoyle over more than four decades. Although Milton Wainwright's stratospheric "biological entities" are not recognizable terrestrial species, they are strikingly similar to Earthly microbiota for the simple reason that life on Earth is indeed derived from the vast — perhaps infinite — reservoir of cosmic life. This represents a philosophy that is alien to contemporary Western thought but is one that has been asserted by many Eastern philosophers centuries before the start of the Common Era. Confucius in China and Sakyamuni (Buddha) in India (ca. 5th century BCE) are foremost among the philosophers whose world view encapsulated the inter-connectedness of all living beings on Earth and throughout the cosmos. It is high time that cultural obstacles to accepting the facts such as are presented in this book are overcome and that we move forward to charting a post-panspermia future for our planet.

Gensuke Tokoro
CEO, Institute for the Study of Panspermia and Astroeconomics
Gifu, Japan

Contents

1

Introduction

We are led to believe that modern science is free of all forms of irrational prejudice that plagued science over the centuries. In this book we document an instance when this is far from true in relation to the most fundamental aspects of biology — the question of the origin of life and its cosmic provenance. From the early 1980's evidence in favour of the theory of cosmic life and a version of panspermia, developed by Fred Hoyle and CW has grown to the point that its continued marginalisation, or even outright rejection, is a cause for serious concern. We present here the story of panspermia in which we ourselves have been directly involved.

From the time of the earliest philosophies of classical Greece the perennial struggle has been to disentangle religion and the "gods" from their involvement in explanations of the external world. Democritus (460–370 BCE) and Epicurus (341–270 BCE) held firmly to rationalist explanations including the concept of an infinite and eternal universe. They both supposed that all matter is comprised of indivisible particles (atoms) and that all phenomena in the natural world — including life — are the result of such atoms moving, swerving, and interacting with each other in empty space in natural and predictable ways. Nothing important was left to mystery or the gods. Although most of Epicurus' writings have not survived into the modern age, a long succession of his disciples have fortunately recorded and transmitted his views, particularly Metrodorus (331–277 BCE), and much later the poet Lucretius (99–55 BCE). The surviving writings of these authors

1

bear testimony to a profoundly post-modern Epicurian view of life in the cosmos. At around 400 BC Metrodorus of Chios wrote thus

> It is unnatural in a large field to have only one shaft of wheat and in the infinite universe only one living world......
>
> (Metrodorus)

> Nothing in the universe is unique and alone, and therefore in other regions there must be other Earths inhabited by different tribes of men and breeds of beasts....
>
> (Lucretius)

Such an evidently post-modern set of ideas relating to life implied also a Universe that was essentially independent of control by any god or pantheon of gods.

The same freedom from theistic control was implied in pre-Socratic ideas relating to the origins of life first attributed to the philosopher Anaxagoras of Clazomenae (500 to 428 BCE). Anaxoragas posited that 'seeds' (*sperma*) are distributed everywhere (*pan*) throughout the cosmos — *pan* linked with *sperma* signifying seeds of life everywhere and thus defining the etymology of the modern word *panspermia*. We should note, however, there are much earlier references to the same basic idea in the wider world outside of Europe. Ancient Egyptian papyri and engravings have references and depictions of panspermia that date before the second millennium BC; and even older Vedic traditions of ancient India encapsulate ideas concerning the cosmic nature, antiquity and eternity of life. Vedic ideas on the antiquity and ubiquity of life found their way into Jain as well as Buddhist philosophy, as for example in this quote from a Buddhist text:

> As far as these suns and moons revolve, shedding their light in space, so far extends the thousand-fold world system. In it there are a thousand suns, a thousand moons, a thousand inhabited Earths and a thousand heavenly bodies. This is called the thousand-fold minor world system...
>
> (*Anguttara Sutta*, c.1st century BCE)

The non-European origin of the concept of panspermia, in our view may well have played a role in the development of the continuing prejudice against it. If this sounds strange to some readers, we need only to recall the violent European resistance to the adoption of decimal number system that we now use everywhere, and on which the whole of modern mathematics and modern science depends.

One of the earliest recorded references to this Indian (Arabic) number system dates back to the mid-seventh century just after the rise of Islam. In the fragment of a document dated 662 CE, Severus Sebokht, a bishop of the monastery located on the Euphrates in Syria expresses his admiration for the Indians because of their method of computation 'done by means of nine signs.' The Indian system seems to have been known in Baghdad as early as 770, or less than a decade after its founding, but it was principally diffused through the writings of the Abbasid mathematician and inventor of Algebra, Al-Khwarazmi, who died around 846.

This so-called Hindu–Arabic number system was firmly rejected throughout Europe for centuries in favour of the bizarrely inconvenient Roman numerals, and it was not until the 16th century CE that the Hindu numerals (renamed Hindu–Arabic numerals) replaced the old Roman numeral system (2). The delay in the transition was undoubtedly connected with a deep-rooted suspicion of an alien non-Christian pagan culture from which this number system had emanated. We leave it to the reader to decide whether the present resistance to accepting evidence favouring panspermia is a cultural response with a similar cause. After all, one of its main modern proponents hails from a distinctly non-European heritage, and one of us MW has his roots in the working class north of England! The late Sir Fred Hoyle also has similar roots and always hinted that this was a major cause for his marginalisation in London-Cambridge based scientific circles.

It is now well over 200 years since Charles Darwin's grandfather, Erasmus Darwin, speculated on evolution. Charles Darwin's *On*

Fig. 1.1. Chandra Wickramasinghe and Milton Wainwright at a UN Space Meeting in Austria.

Origin of Species then later synthesised and further developed ideas that were originated earlier by, amongst others, Lamarck, William Charles Wells, Patrick Matthew, Robert Chambers, and Alfred Russel Wallace. For over two centuries, the ideas of these great thinkers, and the others who followed have loomed large in our exploration of the question of the origin of life and its evolution on our planet. Mostly, if not exclusively, the best-known contributions to the question of life's origin and evolution, at any rate in the modern European tradition has been confined to ideas that are cosily centred on our planet.

In this book we review the history — the alternative, non-Earth-centric, theories of the development of life on our planet. We offer the currently available evidence for the bold view that life on Earth originated from space from an all-pervasive cosmic biosphere (so-called idea of panspermia); and, even more heretically, that life in the form of bacteria and viruses continues to arrive from this source (the theory of neopanspermia).

The philosophical concepts underpinning panspermia have ancient roots that date back to the pre-Socratic Philosopher, Anaxoragas, and even earlier to Egyptian and Vedic traditions. This early history has been dealt with in our earlier publications and so will not be the focus of our present work.

While there has been much speculation about panspermia over the centuries at any rate in the latter part of the common era, there has appeared no defining book or treatise, based on experimental evidence to adequately support its case. Darwin's *On the Origin of Species*, provided such a tome for the promotion of the idea of evolution in general and natural selection in particular. Darwin's great contribution was not in having himself originated the idea of natural selection or evolution more generally, but in providing experimental evidence for its support and by bringing together all the available Victorian ideas on the transmutation of species in a single volume.

Fred Hoyle and one of us (CW) developed the first truly scientific, modern theory of panspermia, combining a vast and rapidly expanding body of astronomical knowledge with a relevant body of biological data such as was available in the late 1970's and 1980's. Since Hoyle's death in 2001, Chandra and his collaborators (including MW) have continued to extend this work (Fig. 1.1). The present book completes this journey by providing experimental evidence for neopanspermia, showing that life must be continually arriving on our planet from space. We also argue that this import of biological information has greatly influenced the evolution of living things on life on Earth and furthermore that the ingress of viruses from space is responsible for both ancient and even up to the minute pandemics. It is our contention that evidence of this point of view grows steadily from diverse fields of science and from many directions.

It is part and parcel of the folklore of the Indian subcontinent that mist and drizzle brings down microorganisms from higher in the atmosphere, microorganisms that sometimes cause disease in plants and animals. We ourselves are now convinced that such rain can indeed wash down a myriad of microscopic life forms from the atmosphere, organisms that arrive in the stratosphere from the furthest depths of space. Strange and unusual life forms, viruses and DNA from space interact with organisms that are already resident here on Earth, and by bringing new genetic information could aid in the evolution of life.

Eons ago this same source of input brought life to Earth at the exact point when it had the conditions that could support biology. Subsequently, repeated waves of new space-derived organisms, often competing with one another, set up the entire pattern and structure of the evolution of life — Darwinian evolution as we now call it.

While we ourselves accept the Darwinian theory of natural selection as a working hypothesis, we are also certain that the development of the evolution of life on Earth is far more complex (not least by the appearance of epigenetics) than is provided by the prevailing view. For example, we do not accept that natural selection on the Earth can explain how complex cells, and the complex hereditary machinery based on RNA and DNA came into existence in the first place. When such ideas were first discussed there appeared to be a comfortable billion years or so of Earth history between the formation of the Earth and the appearance of the first microbial fossils in the geological record. That time interval has now all but vanished, with the earliest evidence of microbial fossils (in the Jack Hills formation) occurring during or very shortly after the condition on the Earth surface permitted life to exist.

Returning to Darwin, it is also worth noting here that he signally failed to fully credit his sources in the first edition of his famous book, and also to provide an adequate history of evolution theories prior to its publication. He claimed that his omissions were due to lack of time and space, but others have suggested darker and more sinister motives. By way of a contrast we provide here a full history of panspermia dating back from the earliest times to the version of this theory forged largely by the work of Fred Hoyle and one of us (CW) in the 1970's and 1980's. There will no doubt be many Darwinists, and biologists in general, who will take a dim view of our findings and our assertions, and like Archbishop Wilberforce will rant and rave as their comfortable world view comes tumbling down!

We cannot stress too strongly that our ideas are firmly based on the well-established methodology of science and upon hypotheses thoroughly worked-out and verified by scientific experiments. All we ask is that our findings be given a fair hearing. The reader might be convinced that we are wrong in our conclusions, but we present them as a sincere attempt to question the current paradigm that appears to us to be woefully flawed and inadequate to cope with an ever-increasing body of new data. Life on this planet, we confidently assert, originated from space, and continues to arrive on Earth from the depths of the cosmos.

2

Microbes and How They Gain a Living

Microorganisms (or microbes) are microscopic organisms, living entities which range from relatively large protozoa, to incredibly minute viruses. Microscopic fungi (i.e. moulds like *Penicillium* species), yeasts and green photosynthetic algae (such as *Chlorella*) also come within the orbit of microbiology. Protozoa, like *Paramecium* were once regarded as being animals, while algae, bacteria and fungi were traditionally studied by botanists.

As a boy, one of us (MW) would spend hours peering down a small German-made microscope (costing one pound fifty — it was about six inches in height and went under the name "The Bijoux Microscope"). The project involved collecting pond water, adding a tiny amount of Oxo and then leaving it until a grey pellicle formed. This was teeming with *Paramecium*, so full in fact that they could barely move. The original microbiologists referred to this organism as the "slipper animalcule".

The Dutch draper, Anton van Leuwenhoek, who is credited with inventing the microscope and who first described seeing bacteria (in 1665) is generally credited with discovering animalcules, or microbes. If you want to split hairs however, Robert Hooke's earlier observation of microscopic fungi (i.e. moulds) was in truth the first microbe-sighting. Hooke went on to popularise microscopy, making it such that any respectable intellectual of the day was also an amateur microscopist.

There is a comedy play, called the *Devil on Two Sticks*, written by the English playwright, Samuel Foote in late seventeen hundreds that refers to animalcules causing disease. Although the play is a satire, it shows that the germ theory of disease was speculated upon then, and perhaps even earlier in that century. Who then first scientifically linked microbes with disease? Most people would reply, Louis Pasteur, but in fact, the Scottish pathologist Sir John Goodsir got there first. While working on stomach illnesses, Goodsir observed bacteria in the stomach contents of a patient, named the organism (*Sarcina venticuli*), went on to show that the organism caused the disease and finally affected a cure by administering creosote. Which brings us nicely, as things often do, to Charles Darwin!

Throughout his later life, Darwin suffered from incurable stomach problems. In desperation, he sent samples of his vomit to Goodsir, who, finding no *Sarcina* in them was unable to suggest an effective treatment. It used to be generally believed that Darwin suffered from Chagas disease which he contracted when on board the *Beagle*. We now suspect however, that he suffered from lactose intolerance, a problem that was not known about in his day. One of the present authors (CW) was in fact diagnosed with the same malady many years ago, and informed that this was far more common in people of Asian rather than European descent.

Before we go any further, we need to have a look at what microbes really are, and discuss how they "gain a living", using a variety of so-called "metabolic strategies". **Aerobes** are obviously organisms which require oxygen, and these are always so-called heterotrophs. **Heterotrophs** gain energy and carbon by using oxygen to oxidise carbon substrates, like sugar and cellulose to carbon dioxide (think of how humans and animals live). **Anaerobes** are the opposite, they are heterotrophs, but grow in the absence of oxygen (they are always bacteria). **Chemoautotrophs** and phototrophs are what are termed autotrophs, that is they are aerobes, but do not use carbon as a substrate for growth and energy. Chemoautotrophs get their carbon by fixing carbon from the air, and their energy from oxidizing a substrate like sulphur or iron (again, these are always bacteria). Lastly, **phototrophs** are photosynthetic organisms that use sunlight to split water in order to gain energy, and fix carbon dioxide from the air for their carbon needs (higher plants, algae and some bacteria are phototrophs). The most efficient way for an organism to grow is to use aerobic heterotrophy, hence humans, who need a huge energy supply to operate their brains and bodies, and elephants needing an elephantine amount of carbon and energy to achieve their size, are both heterotrophs.

We assume that biochemistry operates in the same way throughout the cosmos, so we assume that organisms living "out there" will use the same so-called, "metabolic strategies as described here. Similarly, we assume that the heredity mechanism of life will be based on RNA and DNA throughout (of course we may get a surprise and find other systems operating elsewhere, which are unknown to us). Based on these assumptions, we expect that an intelligent alien will be an aerobic heterotroph and not a bacterium, or plant.

How Might Life on Earth have Originated?

The Earth formed from smaller pieces known as planetisimals in a newly formed solar nebula between 5 and 4.5 billion years ago. The first evidence of microbial residues (in the form of C-12 enhanced residues) are found in rocks dated between 4.2 and 4.3 billion years ago almost immediately after it was possible for microbial life to survive at the Earth's surface.

The standard scientific view of how life began on Earth is well known. On the prebiotic earth, chemicals interacted to give the so-called "building blocks of life", such as amino acids, which then reacted to give the even more complex components of biology, such as DNA and proteins. Essential components of the cell, such as membranes, nuclei and mitochondria were then formed *and life came into being*. Of course, all the true and mysterious complexity of this idea is embraced by the italicised words. This kind of paragraph, and similar unproven assertions, can be found in modern student texts and even more advanced works. It can involve the use of slightly different wording, but this is essentially the same explanation that has been given by scientists for over a hundred years and more. Take for example, this passage by H.J. Hardwicke in his *Evolution and Creation* of 1887:

> The first form of life, says Haeckel, was the Monera, a structure less aluminous, *"atom of bioplasm"*, not even passing the structure of a mere cell. *This primordial organism gradually developed into single nucleated cells.*

Yet again the problem lies with our italicised words. Now, not surprisingly, explanations of the origin of life have become extremely sophisticated over the last fifty, or so, years as our awareness of all kinds of biological complexity has developed, not least with an understanding of the role of nucleic acids (RNA and DNA) in life. Much of the origin of life theory nowadays involves more and more sophisticated chemistry showing that literally thousands of chemicals which are fundamental to biology can be naturally synthesised under diverse conditions, which, it is claimed with little evidence, were involved in the *de novo* origin of life. A whole string of theories has recently been generated to account for life's origins, including, the "Clay Origin of Life", the "Underwater Smokers" origin and the so-called "RNA World".

Did Life begin in Hydrothermal Vents?

Many scientists have suggested that life began under the sea, in what are termed hydrothermal vents, fissures on the seafloor of volcanically active areas which continually discharge geothermally heated water. Hydrothermal vents often, so-called black or white smokers, are extremely biologically productive and are rich in chemosynthetic bacteria (i.e. those growing without a carbon food-source) and primitive archaea. These microscopic organisms act as the base of a food chain which supports the growth of higher organisms, including, clams, limpets, shrimps and impressive giant tube worms. Hydrothermal vents are also thought to be active on Jupiter's moon, Europa and Saturn's moon, Enceladus; they may also once have existed on Mars. Gunter Wachtershauser suggested the so-called iron sulphur-world as an explanation for life's origin in hydrothermal vents. Amino acids may also have been formed in the methane and ammonia-rich vents which were then transferred to cooler waters, where they interacted with clays to form proto-life forms. It has also been suggested that life began in areas of Earth subjected to impact events, such as the Chicxulub impact. Such impacts could have increased the number and variety of hydrothermal vents on the early Earth and therefore increased to number of locations where life could conceivably begin. Impact events like these are also thought to cause the mass extinctions of life that marks the K/T boundary, and are therefore thought to be involved in both the birth and extinction of life on our planet.

The Clay World Hypothesis

A model for the origin of life using clay was first proposed by Alexander Cairns-Smith in 1985. The hypothesis suggests that complex organic molecules arose gradually on the surfaces of silicate crystals in solution. Studies have shown that aqueous solutions of montmorillonite clays catalyse the formation of RNA by inducing nucleotides to form longer chains.

The RNA World

The RNA world hypothesis suggests that life on Earth began with a simple RNA molecule that could copy itself without the help of other molecules. Random RNA sequences have properties useful for the origin of life, but, based on what is known of current life it is not possible to come to the conclusion that a self-replicating molecule was the primary source of the catalytic functions fundamental to living systems. In 1996, NASA concluded that the RNA World hypothesis is plagued by "significant difficulties" notably by the chemical fragility of RNA and the fact that it possesses a very narrow range of catalytic activities. Charles Carter, a structural biologist at the University of North Carolina, has provided evidence that RNA molecules alone would not have been sufficient for all the processes needed to initiate life on Earth, and Harold S. Bernhardt even referred to the idea as 'the worst theory of the early evolution of life', stating that RNA's complexity means it could not have arisen prebiotically.

Life-Particles or Jeewanu

In the 1960s, the Indian scientist, K. Bahadur reported synthesizing living globules called "Jeewanu" (*Jeevanu* is a Sanskrit word meaning particles of life). These, he claimed had similar properties to so-called coacervates, suggested by Oparin. Bahadur believed that life is an inherent property of matter which can appear under special conditions and that Jeewanu are living units. Although the Sri Lankan Chemist Cyril Ponamperuma, was unable to repeat Bahadur's work, the Hungarian chemist Tibor Ganti considered the Jeewanu a promising, overlooked, model system to understand the origin and fundamentals of life. Experimental replication of Jeewanu was published in 2013 by Gupta and Rai who claimed that the globules showed growth, metabolic activities, and the presence of RNA-like material; they further claimed that the RNA-like material detected in the Jeewanu protocells support the RNA world hypothesis. Yet again, we have a theory, based on some experimental data, which warrants further examination and more rigorous scrutiny.

The above theories are intellectually impressive, and satisfying to some, yet they fail to provide a convincing explanation of the origin of life and we are still left with unsatisfactory statements that conclude each of these claims----*and then life began*. All of this begs the question — did life ever originate on Earth? At first sight, this question seems so ridiculous, why would anyone bother to ask it! Perhaps however, life in the cosmos *always was*. Now, this appears like a religious statement but it becomes more explicable when we observe that the human mind is programmed to think in terms of beginnings and ends. We think of life beginning and ending, likewise with degree courses and marriages starting and often ending. We are hard-wired then to think in terms of beginnings and endings. So it is with life — it must have had an origin. This feels like an unsatisfying explanation, but it is similarly impossible for the vast majority of people to accept the concept of infinity (other than by employing abstract ideas of mathematics). We believe infinity, an ancient Indian concept, must be true because we have been told it exists by more sophisticated minds than our own, but still we emotionally want an *ending* to space. Louis Pasteur clearly had doubts about life having an origin as can be gleaned from this statement:

> I have sought spontaneous generation without discovering it for twenty years. No, I do not consider it impossible. But by what authority do you maintain that this is the way life originated? You put matter before life and thus make inanimate life eternal. Who tells you that the incessant progress of science will not compel scientists a century, a millennium, ten millenniums hence…to declare that life and not matter is eternal.

Biologists who claim that life began on Earth need to explain why this would happen on an insignificant planet in the corner of an equally unremarkable galaxy. The answer could be that it was just pure happenstance. Of course another answer might be that it reflects the notion that we believe that we are the "centre" of the universe, looking out from ourselves to the cosmos; life must have begun on Earth we tell ourselves because we are nature's (God's?) ultimate creation. Well maybe we are, and this is the only planet on which life originated and continues to exist. Probability arguments, however, clearly and overwhelmingly suggest otherwise.

Organics from Space and Life's Origin

It is remarkable how science progresses with the changing times. In the late 1980s, the claim by Fred Hoyle and CW that interstellar space is full of organic molecules, derived from comets, was ridiculed; the claimed absence of these compounds (from poor astronomical data) was then used as a stick to beat the theory of cometary panspermia. Now, an internet search readily brings up news, with titles like "The Building Blocks of Life May Have Come from Space" Although this latest evidence is claimed to be a form of panspermia in the true sense it is not, since panspermia (life everywhere) refers to living organisms, rather than chemicals; the presence of organics in space, and the fact that they can be transported to Earth and elsewhere is however, obviously relevant to the origin of life on this planet. As one of us (CW) and Fred Hoyle maintained stridently in the 1980's is that they represent the degradation products of biology, firmly attesting the validity of the theory of cometary panspermia.

In 2005, Jocda Llorca and co-workers, for example, analysed cometary dust in the stratosphere and space (notably from comets) and found a great variety of organic compounds. They conclude:

> Since substantial amounts of cometary dust were gently deposited on Earth, their organic content could have played a role in prebiotic processes prior to the appearance of microorganisms.

They also suggest that the moons of Jupiter may also likely have received such inputs.

Scientists at the University of Massachusetts have also recently found complex primordial organics in cometary nuclei and suggest that radioactive isotopes in comets provide heat to melt cometary ice, providing organic-rich water over "geologically and biologically significant time periods", which could have been delivered to the lifeless Earth; a process which will obviously continue to the present day. It is precisely this model of radioactively heated cometary pools that Hoyle and CW discussed in their 1984 book *Living Comets* not in the context of transporting lifeless organics but as sites for the maintenance of microbiological habitats over billions of years — thus validating panspermia.

A modern version of the iconic Urey–Miller experiment has been conducted by R.J. Kaiser at Berkeley, who combined carbon dioxide, ammonia and other chemicals, like methane, at super cold temperatures and created a "mini-comet". When they exposed this to electrons, amino acids and dipeptides were formed. They conclude that:

> This study reveals unambiguously, that extra-terrestrial polypeptides, once produced, could have inseminated early Earth via comets, and catalysed its biological evolution.

Impact events can, it seems alter, these organics. For example, remarkable, new evidence shows how simple organics may have been transformed into more complex pro-life molecules. In 2015, for example, Harun Sugahara of Nagoya University conducted shock experiments on a frozen mixture of amino acids, water and silicate at cryogenic temperatures (77 K), and a vertical gun was used to simulate cometary impact-events. Significantly, amino acids were converted to proteins. While such results are exciting and vindicate part of the Hoyle and Wickramasinghe Theory of Cometary panspermia, we are still left with the nagging question (at least in relation to the chemical theory of life's origin) — how do these chemicals, however formed, come together to form a living organism? The odds against this remain superastronomical.

Importance of Aerobic Heterotrophy to the Development of Intelligent Life on Earth

Aerobic heterotrophy is the most efficient means by which an organism can gain energy. This metabolic strategy provided the energy to allow simple organisms to move, to become predators, and to evade predation. Predation is the major driving force of evolution, since in the concept of survival of the fittest it leads to the selection process that leads to complexity. Such complexity leads to the large efficient brain of Man (*Homo sapiens*) and has (with the development of an opposable thumb) allowed Man to develop technology, which in turn, one might argue, has led to computers, books, telescopes manned space flight and the exploration of the cosmos; all the result of Man's ability to utilise aerobic heterotrophy. No other metabolic strategy could have provided the energy for driving evolution to intelligent organisms which can split the atom, unravel the nature of DNA, and explore space! A corollary of this observation is that highly developed, intelligent or super intelligent organisms can only arise and evolve on an aerobic planet.

Whether aerobic heterotrophy first arrived in an organism from space via panspermia or developed *in situ* on Earth is immaterial, although its establishment on this planet could only have followed the appearance of the oxygen generating process of photosynthesis. The importance of aerobic heterotrophy lies then in the fact that it is the only growth strategy which can lead to predation and the evolutionary arms race which eventually endowed our planet with intelligence.

Silicon and Life

The idea championed by many science fiction writers that silicon could replace carbon as both an energy source and the building block for complex biomolecules is in our view extremely unlikely, simply because silicon tends to form extremely stable bonds. It is therefore difficult to conceive of silicon being used to build biomass and act as an energy source. However, silicon may have played a role in the establishment of the first bacteria that arose in the cosmos, or arrived to Earth. Life uses whatever is available at hand, so it would not be surprising indeed to find that the second most common element on Earth is used by cells in a more direct way, rather than merely acting as a component of plant cell walls and diatom shells.

3

Historical Overview of Life's Origins

The current scientific consensus tends to distance itself from panspermic ideas and resolutely maintains that life on Earth arose as a one-off act of spontaneous generation (abiognenesis) at some point in the past. This view has a long history; Robert Chambers in his influential, pre-Darwin book *Vestiges of the Natural History of Creation* on transmutation (i.e. evolution) of 1845 stated that:

> The first step in the creation of life on this planet was the chemico-electric operation by which simple germinal vesicles were produced.

While this view has become the orthodox opinion on the origin of life, Alfred Russel Wallace, whose theory of "natural selection" came at about the same time as Darwin's, took the following, different view:

> I submit that.... living protoplasm has never been chemically produced, the assertion that life is due to chemical and mechanical processes alone is quite unjustified. Neither the probability of such an organism nor even its possibility has been supported by anything which can be termed scientific facts or logical reasoning.

The current, popular view, of the likely origin of life then depends firmly on a belief in a single act of spontaneous generation; that simple life arose from organic chemicals here on Earth or from similar chemicals which may conceivably have arrived from space. This proto-life then became more complex, via an evolutionary process explained by Darwinian and neo-Darwinian syntheses.

A largely ignored, schizophrenic turmoil pervades biology in that while most (but not all) scientists believe that a one-off act of spontaneous generation created life on Earth, when conditions differed from what they are at present, these same scientists firmly believe that spontaneous generation is not occurring at the present time. This apparent certainty is largely based on a misreading of the experiments and findings of Louis Pasteur in the 1860s. Andrew Wilson (naturalist and author of the 1877 book, *The Genesis of Life*), summed up the situation as follows:

> There are some who.... whilst assuming that, in the present, life invariably proceeds from life, assent to the statement that in the beginning life arose from non-living matter, they admit the possibility of life-developed *de novo* in the far distant past, but deny the operation of any such action in the present.

It is forcefully claimed that, by using his famous swan neck flasks, Pasteur, once and for all, destroyed the belief in spontaneous generation. In fact, what Pasteur showed was that, if a simple nutrient solution is thoroughly boiled and then exposed to air lacking microbes, it will remain sterile until it is re-exposed to the microbe-laden atmosphere. Pasteur's contemporary critics pointed out that things might have been different had Pasteur changed the conditions of the experiment by using a more complex nutrient solution, by adding some clay or silicon or by incubating his flask over a range of fluctuating temperatures for an indefinite length of time; the possible combinations of conditions that may be argued to have produced life in Pasteur's flasks are, of course, endless. But they are all without any empirical backing.

Pasteur's experiments demonstrated merely that he could kill most microbes by boiling (spores sometimes survived) and that the air is rich in microorganisms which readily contaminate boiled solutions. Although these experiments were essential for the development of the germ theory, the pasteurization of food and the introduction of antiseptic (and later aseptic) surgery, they prove nothing about the existence of spontaneous generation, other than that, if it does occur, it will do so under far more complex conditions than allowed for in Pasteur's experiments. To his credit, Pasteur pointed out this simple fact in the following words, which have been ignored ever since:

> You will note that I never claimed to have proved that spontaneous generation is impossible. In matters of this kind a negative cannot be proved, and: It cannot be proved a priori that there is no spontaneous generation. It is impossible to do more than demonstrate (1) that there are unperceived errors in the experiments, (2) that when the causes of error are removed without disturbing the fundamental conditions of the tests all signs of lower life disappear.

The physicist (who determined the value of absolute zero temperature) and inventor Lord Kelvin was one of the first to suggest that life may have been delivered to Earth by meteors, or comets. Kelvin was not impressed by the arguments of evolution and he claimed to show that the Earth was too young for the process to have occurred. Darwin said of him:

> Thompson's view of the recent age of the world have been for some time one of sorest troubles.

He was also amongst the first to use the terms, intelligent and benevolent design. At the end of his now famous Presidential Address to the Edinburgh Meeting of the British Association in 1871 he stated:

> Because we all confidently believe that there is at present, and have been from time immemorial many worlds of life besides our own, we must regard it as probable in the highest degree that there are countless seed-bearing meteoric stones moving about space.
>
> The hypothesis that some life has actually originated on the Earth through moss-grown fragments from the ruins of another world may seem wild and visionary. All I maintain, is that it is not unscientific and cannot rightly be said to be impossible.

He then points out the paradox between those who maintain that the spontaneous generation of life is impossible, yet apparently arose on Earth once by abiogenesis:

> I confess to being deeply impressed by the evidence put before us by Professor Huxley, and I am ready to adapt, as an article of scientific faith that through all space and time that life proceeds from life and nothing but life.

Kelvin's argument would have probably been made more plausible had he suggested that microbes, rather than moss, were the organisms delivered by panspermia. Not surprisingly, his ideas came in for a lot of criticism from Darwinists. Joseph Hooker, the famous naturalist, friend and champion of Darwin, for example, wrote:

> The notion of introducing life on Meteors is astounding and very unphilosophical... For my part, I would as soon believe in the Phoenix as in the meteoritic import of life.

The possibility then that life may, even at this moment be arising spontaneously in a nearby pond, or perhaps in a deep-sea vent, has never been adequately dismissed (of course if it did so, it would likely be consumed by existing life). The founder of endocrinology, coiner of the word insulin and discoverer of adrenalin, Shapley Schafer said as much in 1924:

> We can by no means be certain that the evolution of non-living into living cannot be happening still.

The idea that life is originating from simple chemicals today goes so much against the current paradigm that it seems impossible to even contemplate; but why then should we accept that it happened in this way in the past? The simplistic answer, is that conditions were different then!

The last serious attempt to demonstrate that life can have a modern origin was by the British pathologist and dedicated evolutionist, Henry Charlton Bastian. Working in the late 1800s and early twentieth century, Bastian believed that life could originate *de novo* (abiogenesis, or archebiosis) as well as from the products of previous life (heterogenesis). His work was never replicated however, and much of it can be explained by the ability of bacteria to grow in apparently nutrient-free solutions, particularly in the presence of added silicon compounds. Bastian believed that bacteria must have been continuously originating based on the fact that they, even over vast periods of time, have never evolved into more complex organisms since life is said to have originated on Earth. He suggested bacteria and other microorganisms arise *de novo* throughout history and that modern forms arose now and have not had time to evolve into more complex organisms. It is interesting to note that Bastian was a confirmed Darwinist and asserted that spontaneous generation was a necessary component of requirement for evolutionary theory. Darwin read Bastian's work with interest, and concluded that while Bastian had some good ideas, he did not accept his findings; he remained thoroughly convinced of the truth of archebiosis however. Thomas Huxley, "Darwin's Bulldog" also asserted that "if the theory of evolution is true, then living must have arisen from non-living".

Huxley made his own contribution to the origin of life theory when he claimed to have discovered, *Bathybius haeckelii*, a substance that he, at first, believed was a type of primordial matter and the source of all organic life. It was however, later shown to be the product of marine precipitation; surprisingly, Huxley was even accused of deliberate fraud on the matter. In 1868 Huxley examined an old sample of mud from the Atlantic seafloor and thought he had discovered an organic substance which he called *Bathybius haeckelii* in honour of the German biologist, Ernst Haekel, who had theorized that life originated from a substance he called "Urschleim". Huxley thought that *Bathybius* could be that link between dead and living matter. In 1872, the *Challenger* expedition spent three years studying the world's oceans, but found no evidence of *Bathybius*. Huxley recanted immediately, but Haekel believed in *Bathybius* until 1883.

However, not everyone has taken this materialist, Earth-centric view of life. Alfred Russel Wallace for example, believed that if life is unique to Earth and that it exists everywhere (particularly intelligent life) then this reality would merely reduce Man to an animal and of no particular importance in the Universe. He believed that Man's intellect may well have been directed by some higher intelligence. Contrast this with the following statement, again by Thomas Huxley, which is in complete agreement with the generally held modern view of life and how it came into being:

> A mass of living protoplasm is simply a molecular machine of greet complexity, the total results of the works of which, or its vital phenomenon, depend, on the one hand, on its constitution, and on the other, upon the energy supplied to it; and to speak of "vitality" as anything but the name of series of operations is as if one should talk of the horology of a clock.

The modern microbiologists, Woese and Goldenfield sum up the current status of biology brilliantly, as follows:

> Biology.... is a discipline whose perspective is that of classical 19th-century physics, inherently incapable of dealings with the problems of a nonlinear world, which is non-reductionist, non-deterministic (acausal), and works in terms of fields and emergent properties, not a static world of particles with linear relationships among them.

Andrew Crosse and an Early Attempt to Create Life

A fascinating story claims to link Mary Shelley's Frankenstein and the history of theories on the origin of life. Frankenstein or the modern Prometheus was published in 1818 and it has been suggested that Shelley got the idea of creating life from the awareness of the work of a certain Andrew Crosse, a claim which, in fact, turns out to be wrong. Crosse's story remains interesting however. He was regarded as something of an ecentric and referred to locally as the "The thunder and lightning man" because of his experiments conducted in the Quantock Hills of Somerset in England (Fig. 3.1).

Andrew Crosse claimed he could create life in the laboratory merely by the passing of low electric current through a solution of potassium silicate. The life forms he claimed to produce were small mites (which he called *Ascarus electricus*, or, *Ascarus crossii*) with fine hairs, which after twenty-eight days he said, began to move. The

Fig. 3.1. Andrew Crosse, "The thunder and lightning man".

experiments were, it seems repeated by his friend W.H. Weeks. Crosse never claimed that he had created life *de novo*, believing instead that the electric current probably stimulated some air borne ova into hatching; despite this, he claimed the same result occurred when the electrified solution was kept sealed from the air. The experiments were later unsuccessfully repeated by a German Professor, called Schulze. In 1837, a number of newspapers reported that Michael Faraday had also replicated Crosse's results, and discussed them at a meeting of the Royal Institution on February, 28th, 1837. However, this appears not to have been the case; Faraday it seems, never even attempted the experiments. Incidentally, Faraday made an interesting comment about life in space at the Royal Institution in 1838:

> In other planets, other beings may and very probably exist, and all things agree wonderfully suited to them as they are with us; but as yet we do not possess sufficient observations to decide the question.

In the early nineteenth century, Andrew Crosse lived in an isolated country house in the Quantock hills of Somerset. Based on his experiments on electricity, he was locally known as the "Wizard of the Quantocks the thunder and lightning man", or simply, "the electrician". The experiments were conducted around 1805 in a laboratory set up in his home, Fyn Court. Mary Shelley' diary reveals that she, and her husband, Percy Bythe Shelley attended a lecture given by Crosse (Fig. 3.2) in London, in 1814. It seems likely that the Crosse's lecture gave Mary Shelley the idea of using electricity to breathe life into Frankenstein's monster.

Crosse's work had an unfortunate impact on the history of evolution. It was quoted by Robert Chambers in his *Vestiges of the Natural History of Creation* as a valid scientific experiment which provided proof of spontaneous generation. Religiously biased critics of transmutation (i.e. evolution) could point to the obvious ridiculousness of Crosse findings, and summarily dismissed Chambers', otherwise rational ideas on evolution.

Darwin and the Origin of Life

Charles Darwin (1809–1882) lived through a period of history that saw two revolutions in biology. The first encouraged by his own work was the revolution in evolution, the second was the growing understanding of the role that microorganisms, or germs, play in human diseases. One might get the impression from reading biographies of Darwin that this second revolution passed him by totally, and this was the impression one of us (MW) got before delving more deeply into the connections between Darwin and germs. While Darwin worked on such diverse topics as insectivorous plants and earthworms, he surprisingly, never worked with microbes or became involved in the controversy raging from 1860 for the next forty, or so years over so-called spontaneous generation as we have seen. This was a long-held idea that living things just arrive from nowhere, or *de novo* (from new) to use the scientific expression. Whether Darwin thought all this new-fangled work on germs was for younger men, or that the new science of microbiology was a somewhat esoteric area of biology, best left to the experts, we will never know. Darwin certainly had a microscope which he was fond of using in his research. With his considerable talents, he might have made a major impact on the development of the role of germs in the environment and as causal agents of disease. Likewise, there is no mention of microbes in the later editions of the *Origin of Species*, and Darwin never introduced bacteria into his thinking on evolution.

Now, it is generally assumed, though yet to be demonstrated, that life arose on Earth from chemical building blocks (recent studies claim that these building blocks themselves may have originated from space), and the first organisms would have been simple and ultimately evolved to bacteria. Darwin however, never dwelled on the origin of life on Earth, thinking it a topic beyond science of the time.

In an 1859 letter Darwin did however, explain to Lyell

> The parent monad-form might perfectly well survive unaltered & fitted for its simple conditions, whilst the offspring of this very monad might become fitted for more complex conditions. The one primordial prototype of all living and extinct creatures may it is possible be now alive!

Darwin later wrote to Thomas Henry Huxley "we know nothing as yet [of] how life originates". Then, in the 3rd edition of *Origin of Species* (1861) Darwin added the remark:

> Science as yet throws no light on the far higher problem of the essence or origin of life.

But perhaps most revealing of all about Darwin's view on the beginning of life is the often-quoted comment from an 1871 letter to Hooker:

> It is often said that all the conditions for the first production of a living organism are now present, which could never have been present. But

if (and oh! what a big if!) we could conceive some warm little pond, with all sorts of ammonia and phosphoric salts, light, heat, electricity, etc, present, that a protéine compound was chemically formed ready to undergo still more complex changes, at the present day such matter would be instantly devoured or absorbed, which would not have been the case before living creatures were formed.

Even to the end of his life Darwin maintained that there was no evidence to support spontaneous generation and no valid explanation of how life originated.

Though no evidence worth anything has as yet, in our opinion, been advanced in favour of a living being, being developed from inorganic matter, yet we cannot avoid believing that this will be proved some day in accordance with the law of continuity.

These two comments are remarkable for 1871; in fact, the first is one of Darwin's most original statements. Darwin would say little about the origin of life, clearly believing that the subject lay outside the orbit of science and could not be subjected to experimental verification, certainly not during the Victorian age. Surprisingly, considering that Darwin is held in such high esteem, this short paragraph (beginning, "it is often said") has been known, but not emphasised. Let us then comment on what we think is remarkable about it. The paragraph sums up the so-called chemical theory of life, the idea that life arose from inorganic chemicals. Surprisingly, life is not something that is easy to define. We all know what life is, we see it and experience it every day, but to sum up in a sentence what distinguishes you the reader from the desk or coffee table in front of you is not as easy as you might imagine. As a result, life is defined by a number of characteristics. Living organisms for example reproduce, they have ordered chemistry that leads to the synthesis of new body material; they can by various means, obtain energy to keep them alive, moving and growing; they use DNA and RNA as information systems; and finally, they generally react often to stimuli. All life on earth is based on carbon, a fact which is no accident, since the chemistry of carbon is superbly fitted to building complex polymers, like proteins which make up the bodies, or biomass, of living organisms. Carbon also readily forms bonds with itself and other elements; these bonds can however, be readily broken to release energy. This means that carbon can be built up during life and decay after death contributing to what is called the carbon cycle. All of us will one day return to the biosphere, the elements of which we contain will then be recycled into new organisms.

Let us get back to Darwin's famous origin of life paragraph. For the moment, let us ignore the first line and go into the paragraph at the point of... But if.... Here we see Darwin describing the essentials of the chemical theory .... some warm little pond before life began in which all sorts of chemicals come together and up pops life; the other essential ingredients apparently being light heat and electricity. Darwin considers here that some form of energy would have to be infused into a system to create life. Subliminally, he is perhaps thinking of the necessity of applying a life force to the chemical mix in order to create life. Modern scientists do not believe in a life force. We think instead that a gradient from non-living to life occurred without the need of a specific life-giving force. Such a life-force has of course traditionally been held by philosophers to exist. Darwin's words can be seen from another angle, what if light and electricity, rather than providing a life force, actually modified the chemicals involved, so that they came together to from the building blocks of life. This is the basis of the experiments by Harold Urey and Stanley Miller on which the modern chemical theory of life is based. In the early 1950s the young Stanley Miller was looking for a research project to work on for a PhD with his supervisor Harold Urey, already a Nobel Prize winner, who had formulated a program aimed at attempting to make the basic chemical building blocks of life, such as the amino acids, from inorganic chemicals.

The task before Miller was therefore to create the chemical building blocks of life using inorganic chemicals and conditions that might have been present on prebiotic Earth. The apparatus used was relatively simple, but potentially explosive! A mixture of gasses contained in a glass vessel could be exposed to spark and the resultant products dissolved and condensed in water. When Miller passed a spark through a mixture of gases, to everyone's great delight, he produced a mix of amino acids. Since then, most of the important building blocks of life have been formed using this and other similar methods. However, controversy still reigns over whether the conditions used actually simulate the prebiotic earth, but such arguments are splitting hairs because the fundamental fact about these experiments is that they tell us a lot about chemistry, but very little about the origin of life. The best that can be said is that it is easy to make the relatively simple building blocks of life, but amino acids or any other chemical mixes do not form living organisms while sitting there in reagent bottles on laboratory shelves! Biologists have hyped these experiments beyond belief! Their latest manifestation is in the probability that most of life's building blocks were delivered to Earth from space in, for example comets, a kind of chemical panspermia. It doesn't matter however, where these chemicals come from, there remains an infinite distance between them being present and life beginning. Darwin assumes in his famous paragraph that life would originate in the presence of light. Light however, is germicidal and would likely have denatured any proto-life forms. Interestingly most traditions suggest that life began in a dark watery world.

There is a rare pamphlet that was published in 1894 by the Leeds Astronomical Society. In this, which is the second of the Society's Reports and Transactions, can be found an article by the Society's then president, a certain Mr W.D. Barbour. Mr Barbour discusses the possibility that life exists elsewhere in the universe and makes the following relevant quote on the early ideas on the origin of life.

> Other ancient traditions, or cosmogonies, are corroborative of the "dark water production" of life. Chaldean represents water as the producing mother. Babylonian commences with "nothing but darkness and an abyss of water". An Egyptian story of the origin of

the universe, mentions an abyss enveloped in boundless darkness. Phoenician alludes to an eternal dark chaos and creatures developed from water.

These ideas are reflected in modern scientific ideas that life may have first begun under water smokers, a nice enough idea, but there is absolutely no evidence to support it!

Microbial invasion from Venus!

Unusual, somewhat speculative evidence for panspermia was provided by the astronomer, D.R. Barber whilst working at the Norman Lockyer observatory near Sidmouth in England where he spent much of his time developing photographic plates. The procedure for developing plates in the old days involved aqueous solutions of various silver halides. Over a twenty-five-year period between 1937 and 1961 Barber observed six unique events involving the mysterious fogging of his photographic plates. This was caused by the appearance in water (from an isolated well-supply) of large numbers of gelatine-liquefying microbes. These caused major damage to photographic material in the brief period (three quarters to one hour) between the washing and drying processes of astronomical plates and films. The growth caused distinct pitting of gelatine silver layers of 0.05 to 0.25 cm. The presence of microbes in the pits was first detected using a stain called methylene blue, which is widely used to stain bacteria. Over thirty photographic astronomy records were destroyed. Then in 1956 there was similar invasion of airborne microorganisms. The source of the pitting was laid at the door of two organisms, an unidentified yeast and the bacterium, *Bacillus fluorescens*. Both organisms were silver resistant. Barber stated:

The results showed that each recorded presence of abnormal quantities of microorganisms in the local (supply well) or on processed photographic marker coincided closely in date with an inferior conjunction of Venus and a major geomagnetic storm.

So why has no one else seemingly reported this event? Well, they have suffered the same problem, but not being astronomers failed to recognise its significance. Barber also said that he employed a special processing system which may have enhanced the phenomenon. He also pointed out that the occurrence coincided with northerly winds which apparently would have been expected to have brought particles from the upper atmosphere to that part of Devon. So an experienced observational astronomer links microbial influx with the transit of Venus. Barber, however, may have been over-optimistic in assuming microbial transfers with every inferior conjunction of Venus; realistically the two planets would be connected biologically over timescales of several decades. Nevertheless, the next transit of Venus is in December 2025. We could be there, ready with our Petri dishes containing gelatine and silver ions to test this claim.

4

History of Ideas About Life on Other Planets

> It is unnatural in a large field to have only one shaft of wheat and in the infinite universe only one living world.........
>
> Metrodorus of Chios, (400 BCE)

By the early 1800s many thinkers considered it an obvious possibility that life exists on other planets (spheres and globes, as they were often called). This belief was based largely on the simple dictum that the Creator (generally used in reference to the Christian God) would never have established such a large number of worlds only to leave them empty; or as Isaac Taylor states, in his *Physical Theory of Another Life* of 1836:

> None could tolerate the idea, and especially seeing what we see in our own planet, that the innumerable spheres around us are totally untenanted, and that the stupendous celestial mechanism, is a mechanism merely.

A particularly common feature of this early literature is how often a view is given that living things are *adapted* to the conditions found on Earth and that such adaptation occurs in life-forms which exist on other planets. Such ideas were current before Charles Darwin wrote his seminal work on evolution in 1859.

The *Vestiges of the Natural History of Creation* (which predated *On the Origin of Species*), was first published anonymously, by

Robert Chambers (of the encyclopaedia-publishing family, Fig. 4.1). Chambers, unlike Darwin, speculates on the possibility of evolution (or transmutation as it was then called) on other planets:

> The whole train of animated beings, from the simplest and oldest up to the highest and most recent are, then, to be regarded as a series of advances on the principle of development which have depended upon external physical circumstances to which the resulting animals are appropriate... I contemplate the whole phenomenon as being driven in the first place arranged in the councils of the Divine Wisdom, to take place, not only upon this sphere (planet) but upon all the others in space, under necessary conditions and modifications, and being carried on, from the first to the last, here and elsewhere, under immediate favour of creative will or energy.

Chambers went even further, suggesting that:

> Where there is light there will be eyes, and these in other spheres, will be the same in all respects, as the eyes of tellurian animals, with only differences as may be necessary to accord with minor peculiarities of condition and situation.

Fig. 4.1. Robert Chambers. One of number of authors who influenced Darwin.

And finally:

> Thus, as one set of laws produced all orbs and their motions and geognostic arrangements, on one set of laws overspread them all with life. The whole production or creative arrangements are therefore in perfect unity. It is likely...that the inhabitants of all other globes of space bear not only a general, but a particular resemblance to those of our own.

Darwin generally avoided philosophical arguments. *On the Origin of the Species* provides a straightforward synthesis of then available arguments on the species question, extended by various references to Darwin's own observations. He generally avoided speculating at length on issues such as how life originated. In an 1863 letter to Hooker he did however comment, "It is mere rubbish, thinking at present of the origin of life, one might as well think of the origin of matter". Other Victorian thinkers did, however, consider the possibility of life on other worlds, the most famous contribution being, The *Plurality of Worlds* by William Whewell, in 1855. Whewell, like a number of others, speculated on the possibility that the development of life on Earth had been influenced by the catastrophic arrival of comets or asteroids from space, a possibility, which Darwin, the strict gradualist, did not entertain in his writings. Such gradualism conflicts with modern ideas on "punctuated equilibrium", the idea that evolution occurs in spurts instead of following the slow, but steady path that Darwin suggested.

The concept of the "habitable zone" is currently very much in vogue. Sometimes referred to as the *Goldilocks Theory*, it suggests that the Earth is perfectly positioned for its role as an abode for life; as the following quote shows however, this is far from being a novel view:

> Our position in the solar system is very truly affirmed by astronomers to be an extremely favourable one. Less distant from the sun than Saturn, Mars, and Jupiter, and yet unlike Venus and Mars not so near as to feel his power, too — the Earth seems to be fitted for the residence of man during his state of probation.

The above comment was made by William Pinnock in 1836. Continuing with the theme that the majority of current ideas about life have an earlier origin, it is interesting to consider Gaia, the claimed modern idea that the Earth acts like a self-regulating living organism, a hypothesis generally credited to Sir James Lovelock. However, the Victorian geographer, Sir Richard Strachey, came up with the Gaia hypothesis in his address to Geographical Section of the British Association in Bristol on August 26th, 1875:

> The picture I have thus attempted to draw presents to us our earth carrying with it, or receiving from the sun or other external bodies, as it travels through celestial space, all the materials and all the forces by help of which are fashioned whatever we see upon it. We may liken it to a great complex living organism, having an inert substratum of inorganic matter on which are formed many separate organized centres of life, but all bound up together by a common law of existence, each individual part depending on those around it, and on past conditions as a whole. Science is the study of the relations of the several parts of this organism one to another and of the parts to the whole.

Although Strachey's views were further developed by Lovelock, here we have an earlier view of the Earth as an organism, given some hundred years before the idea was formalized as "Gaia". In a similar vein, the Polish-German Professor Eduard Strassberger, (the discoverer of mitosis, cytoplasm and nucleoplasm), rather than emphasising the "survival of the fittest, emphasised the ways in which organisms cooperate with one another:

> But since the life of every living organism on Earth is more or less conditional by the lives of all other organisms, it is quite reasonable from a higher standpoint to regard life on Earth as great *symbiose*.

It is possible then rather than always thinking of organisms out-competing one another, to consider how different organisms arriving on Earth from space might collaborate in the process of evolution.

Since it has always been assumed that life originated on Earth (if it did so) only once, the possibility that life may have arisen multiple times appears novel and exciting. However, if we had a full history of astrobiology it would become obvious that this idea is far from new; Alfred Russel Wallace, for example, stated in 1890 that:

> Now it is manifest that if we look back, as far as possible, into the remote past when the first germ of animal life appeared upon the globe, two conditions of things and two only, are conceivable. Either (A) there was a single germ of life from which all subsequent living forms have been evolved or developed or (B) there are several or many germs of life from which in separate streams, so to speak, the evolution of living creatures took place. Darwin, inclined, I think to the latter supposition; but either A or B must be accepted by all evolutionists of all schools.

In 1865, the German physician, Hermann Richter responded to Darwin's silence on the origin of life by stating:

> We regard the existence or organic life in the universe as eternal and cosmic; it has always existed and has propagated itself in uninterrupted succession.

He termed this extraterrestrial life *cozmozoa*, and suggested that the idea that life comes from space to be logically consistent with other ideas in science and not only that, it provided a cornerstone for Darwin's Theory. Then in 1908, the Swedish physical chemist Svante Arrhenius suggested a novel version of Richter's cozmozoa theory, naming it "Panspermia". Arrhenius provided a mechanism by which life could be transported between planets, suggesting that bacterial spores were propelled through inter-planetary space by radiation pressure.

Although most histories of panspermia are restricted to a discussion of the contributions made by scientists like Richter, Helmholtz, Lord Kelvin and Svante Arrhenius, the subject has a more interesting and complex origin. Who would believe, for example, that the famous inventor Thomas Alva Edison was a convinced panspermist? — arguing that:

> The thought that life originated on this insignificantly little and comparatively unimportant sphere to me seems so inconceivably egotistical.

Edison came up with the idea of what he termed "life units" and added the supposed existence of these to his view of panspermia and came out with this remarkable comment which we used to head the Introduction to this book:

> After the earth cooled of the great heat of its assemblage, life units came to it through space, into which they had been thrown from some other more developed sphere or spheres, Reaching the earth, they adapted themselves to the environment they found here; and then began the evolution of the various species as we have them each growing individual being a collection of cell communes... It is not impossible that, when we find the ultimate unit of life, we shall learn that the journey through far space never could harm it and that there is very little that could stop it...Such units of life have come, and possibly still are coming, without injury through the cold of space.

Comets as the Source of Panspermic Life

The main theory to be developed in this book is the Theory of Cometary Panspermia, that is that comets provide the environment where life exists, replicates and is distributed throughout the cosmos (although asteroids also play a role). Comets have historically been seen not only as harbingers of pestilence and disaster, but also the bringers of life to Earth. For example, the following two quotes from the French astronomer Camille Flammarion, in his influential book, *A General Description of the Heavens* (1895):

> Whence came the first seed, the first germ of terrestrial life? It was either a spontaneous generation, or it descended from the sky.... In the second case, the meteorites which follow the cometary train may come from distant worlds, the debris of which they transport through

space, among which latent germs may survive, ready to fall on a prepared Earth and bloom into the conditions of existence. Comets — their importance would be much greater still if they should be found to carry in them the first combinations of carbon, for it is probable that it was by these combinations that vegetable and animal life commenced on the earth and the other planets and these vagrant bodies might be the source of life on all the worlds.

Mary Somerville, author of *Mechanisms of the Heavens* (1831), on *Molecular and Microscopical Science* and *On the Connections of the Physical Sciences* remarked in 1855 that:

> It has often been imagined that the tails of comets have infused new substances into the atmosphere.

While William Whiston concluded that comets delivered the Flood and that:

> At the worst, comets may be the water carriers for every observation tends to prove that they are masses of transparent fluids.

By the way, Mary Somerville's book *On the Connection*, was the bestselling science book in the UK, prior to the appearance of Darwin's *On the Origin of Species*.

Finally, here is an anonymous quote from the popular magazine, *The Living Age*, of 1861, which links comets and evolution of life on Earth.

> Had we been disciples of Lamarck, we should certainly have fixed on the embrace of comets as the most probable explanation of the development of species, that raising of one type of existence into a higher.

De Maillet: Arguably the First "Modern" Proponent of Panspermia

While Greek philosophers, like Anaxagoras, suggested the possibility that life on Earth arrived from space or other planets in the 5th century

Fig. 4.2. De Maillet: arguably the first "modern" proponent of Panspermia.

BCE, these ideas lay dormant for nearly 2,000 years, until the French philosopher, Benoit De Maillet provided a "modern" (opposed to "ancient") argument in favour of panspermia. De Maillet, a French Government official, was widely travelled and was the author of an anonymous book entitled *Telliamed* (De Maillet spelt backwards). (Fig. 4.2). This book, which featured an imagined conversation between a fictitious oriental (Indian) and equally fictitious European, became a highly influential early text on geology and even detailed a theory of evolution. For many years, *Telliamed* circulated in manuscript form amongst French intellectuals, and was only published in 1748, ten years after De Maillet died. An English edition was published in 1750 although the first unabridged version was only made available in the late 1960s by Carozzi.

The references made by De Maillet to panspermia can be found in a marginal note to his Third Conversation in which he states that "seeds" are found throughout the cosmos; but what does he mean by "seeds"? When De Maillet continues, we see he is not referring to plant seeds, but is arguing that all life forms have arisen from microscopic proto-forms which contain the germ of life, i.e., "seeds", which are impossible to see, even under the best microscopes.

The word "seed" then is used in its broad sense to mean a source, or beginning.

It is also interesting that he speculates that such living forms might be sub-microscopic and visualises that his "seeds" are concentrated around planets and held there by gravity. Further, he visualizes these seeds as traveling across the "void" of space.

Now at this point we might imagine that De Maillet is merely talking about proto-life forms that merely resemble small seeds or sperm, but it becomes clear however, is that he is aware that animalculae (e.g., algae, bacteria and protozoa), are present in water in which grass is present. De Maillet clearly believes that such microscopic forms are the precursors of animals, even humans, which grow directly into the animals and plants we see all around us on Earth. These seeds he considers come initially from the air, but ultimately from space and suggests that the cosmos is full of the "seeds" of all living things and these are transmitted through the universe, becoming more concentrated around rocky planets due to gravity. Once they arrive on a planet with oceans, these worlds, like a vast womb, provide a beneficial climate which nurtures the seeds until they turn into small life forms observable under the microscope. These small animals and plants then grow and eventually leave the sea to colonise the land. Here, they grow to the degree of complexity we see today. De Maillet is aware however, that some living forms do not survive and his reference to evolution induces him to disagree with the views of Sorel, who had claimed that "The species of animals have existed for all times, or have been created in the past".

De Maillet also believed that it possible that all living things existing on Earth occur on other globes (i.e. planets). Finally, he states:

> Furthermore, all the seeds do not occur around every globe. They can only be transferred from one to another by one of these revolutions mentioned previously, to which the entire system is subjected, that is, the passage on globe from one vortex to another. If we could only see all of the species populating our planets, I am convinced that we would discover in them thousands of unknown ones.

At first sight, De Maillet's views on panspermia do not correspond with most recent versions which essentially view microorganisms as the living form which is transported around the cosmos. In the recent version of panspermia, these microbes then go on to evolve into more developed life forms, which, on Earth, has resulted in the form of humans. De Maillet, in contrast, views the cosmos as being full of planets which are populated with diverse plants and animals many of which never evolved on Earth. He also believes the small forms (or "seeds") which arrive on Earth and other planets; develop at first in the oceans and then move to the land; here they then develop directly into their mature forms, some of which fall by the wayside as they do so.

Although the modern versions of panspermia theory generally see microorganisms as the form that is transported throughout the cosmos, it is interesting to note that some other authors have speculated on the possibility that more advanced life forms may be involved in, and may have directed panspermia. Lord Kelvin, for example, as we have seen, talked of massy stones being transported, while Hoyle and CW have even speculated on the possibility that insects may play a role in panspermia. More recently, Tepfer has suggested that plant seeds might also be transported across the cosmos. Echoing yet another theme of De Maillet, some recent authors have argued that a vast array of diverse life may have evolved on other planets, that life on Earth represent only a small sample of life's possibilities, and that these "seeds" contain the genetic instructions for specific life forms.

Thus, it is clear that De Maillet's views are generally consistent with our own more recent ideas based on a more substantial body of evidence. It is interesting to note that *Tellaimed* (De Maillet's treatise) was initially left unpublished because of the fear of a violent reaction these ideas might evoke, especially in ecclesiastical circles. De Maillet was perhaps wise to do so, since, in addition to believing that life was not created by God, he also claimed that it did not even originate here on Earth, but came from space. Such views were unlikely to make him popular amongst a philosophical and a scientific elite most of whom firmly believed in God and his works. De Maillet in fact goes out of his way to avoid any mention of God, although he makes frequent reference to the Bible as providing a means of gaining insight into the mind-set of ancient peoples. Perhaps he was wise to do so considering the fate that befell Giordano Bruno in 1600.

The sad story of Bruno is worth recalling. In 1584, Giordano Bruno a Dominican priest, published his book "Dell Infinito, universo e mondi" ("Of Infinity, the Universe, and the World"). Without benefit of a telescope, Bruno wrote that the stars were just like our sun, that planets must orbit these suns and that intelligent beings, just like the humans of Earth, lived on these planets. Bruno believed life was everywhere and that Earth was not the centre of the biological universe. He was arrested by the Catholic Inquisition, found guilty of heresy, and was then tortured and burned at the stake on February 19, 1600.

De Maillet can be seen as one of the first "moderns" to see the Earth, not as the centre of creation, but merely one trivial infinitesimal part of a vast cosmic sea of life. In this regard, he distanced himself from subsequent philosophers and evolutionists, including Darwin, who believed life originated on Earth. In addition, De Maillet differed from other philosopher-scientists of his age (including Buffon) when he stated that the Earth was two billion, rather than only a few thousand, years old. By becoming one of the first persons to express what might be called a post-Copernican view of biology, De Maillet's contribution to the history of astrobiology and evolution deserves to be better known. Modern commentators keep coming up with ideas about panspermia and the origin of life which they assume are novel, but which in fact turn out to have been voiced, at the very least, hundreds of years ago. It has become fashionable, for example, to talk about the building blocks of life being delivered from space in the form of organic chemicals, but here are a couple of historical views from 1883 and 1872 respectively to add to the ones already given:

> The two or three primordial organic carbohydrates needful to life originated from certain inorganic carbohydrates of the Earth, and these, from interplanetary space. AND:

> Upwards of a hundred meteor systems have provided their tribute of matter from outer space upon this globe, not only during the period of Man's existence, but doubtless from ages before he appeared on Earth. If we not have the seeds of life conveyed to Earth from interplanetary and interstellar space, we have a supply of many materials to support life.

In a similar way, the idea of reverse panspermia is becoming increasingly voiced; here it is stated in the magazine, *Popular Science* of 1922:

> Conversely, germ laden dust is being continuously transplanted from our planet to other worlds. And so this interchange of matter may be eternally spreading the same species of life throughout the universe.

Even the idea of pathospermia — diseases from space — has been aired previously; this example, from the *Report of the Natural History and Philosophical Society of Belfast*, 1889–90 is reassuring as to the dangers of such imported pathogens:

> With regard to the bacilli supposed to be imported from planetary space, if the rules of diminution of strength of a virus (*i.e., the old use of the term to mean an infective agent*) by cold is applicable to them they should be perfectly.

Such ideas are recycled by each successive generation of scientists because they are based on simple speculation, without evidence. In contrast, we have broken this cycle by providing evidence that life and organic materials are continually oncoming to Earth from space. There has similarly been much discussion, ancient and modern, concerning the idea that epidemics arrived on Earth from space. In contrast to idle speculation, Fred Hoyle and CW, in their book *Diseases from Space*, provided evidence, rather than mere theories to show that influenza can be spread in this way.

5

The Astronomy of Panspermia

In the year 2022 one could scarcely imagine that a purely astronomical research project started in 1962 led to a fundamental
challenge to the received wisdom of how we came to be! How did
life originate and come to be dispersed throughout the cosmos,
planet Earth being just one of trillions of similar locales on which
this same life may have taken root and developed over the past
4 billion years. A purely astronomical investigation unconnected with
life into nature's cosmic dust, the dust clouds that show up as dark
patches and striations against the background of stars in the Milky
Way, ended up with the conclusion that interstellar dust must include
vast quantities of bacteria and viruses in various stages of degradation
and decay. These conclusions are still being vigorously disputed by
those who still insist on preserving the *status quo* against all the odds.
The received wisdom in 2022, as we have already discussed, is that life
started *de novo* on Earth from organic molecules that were formed on
an ancient Earth by purely non-living processes. In spite of a growing
body of evidence that challenges this orthodoxy, an institutional and
cultural imperative to maintain the *status quo* remains implacably
strong, and still continues to dominate.

Carbon Dust in the Cosmos

The challenge of the Aristotelean concept of spontaneous generation,
for one of the present authors (CW), began in a most unlikely context

in 1961. CW had started a PhD project at Cambridge to re-examine the composition and sizes of cosmic dust, the vast accumulations of obscuring material, seen as dark markings, shapes and striations against the background of stars in the Milky Way (Fig. 5.1). It is this project that eventually led to the emergence of the ideas of cosmic biology that we shall discuss later in this book.

Fig. 5.1. Clouds of obscuring dust in the mid-plane of the Milky Way.

A prelude to the nature of cosmic dust and the cosmic origins of life, however, goes back earlier to the late 1940's and 1950's and to a monumental program of work undertaken by Fred Hoyle, Geoffrey Burbidge, Margaret Burbidge and Willy Fowler to understand how all the chemical elements were synthesised in stars starting from primordial hydrogen and helium. The conversion of the primordial element H to He was already understood to take place in the deep interiors of stars where temperatures and pressures are so high that nucleosynthesis can take place. The reaction of 4 H nuclei merging in these conditions into He is shown in Fig. 5.2.

It is this nuclear reaction that would keep the sun shining for 7 billion years.

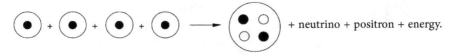

Fig. 5.2. Helium synthesis in stellar cores.

The next step in the process of "nucleosynthesis" in stars involved a prediction by Fred Hoyle in 1949 that if carbon, as we see in all life around us, is to exist in quantity throughout the universe it must also be synthesised in the deep interiors of stars and expelled into space. The relevant nuclear reactions involve a two-step process (Fig. 5.3).

The first step in the process is a reaction of two helium-4 nuclei to produce the element beryllium-8 which is highly unstable, and decays back into smaller nuclei in far less than the blink of an eye — its half-life being less than a billion billionth of second. It is in the requirement of the addition of another helium nucleus to form carbon-12 within this "blink of the eye" that led Fred Hoyle to make his historic prediction of the existence of a hitherto unknown and precisely defined "excited state" in the carbon nucleus, without which carbon could not be produced. This state (now known as the Hoyle state) was not known to exist at the time. Fred argued from his calculations that if such a state did not exist, carbon could not be produced in stars, and consequently we (life) would not exist. This was the first expression of the so-called anthropic principle which asserts that certain properties of the universe must prevail simply because we (humans) are here! In the case of the element carbon, this so-called excited state of the carbon nucleus (the Hoyle state) was later discovered by William A. Fowler at the Kellogg Radiation Laboratory at Caltech, a discovery for which Fowler was later awarded the Nobel Prize for Physics in 1983.

The implication of the prediction of this Hoyle state and its later verification led to the classic publication by Geoff Burbidge, Margaret Burbidge, Willy Fowler and Fred Hoyle (B^{2}FH paper) which laid the foundation for our present understanding of how all the elements in

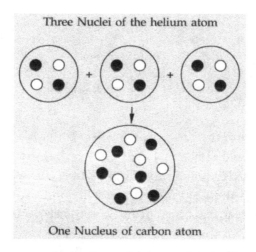

Fig. 5.3. The formation of carbon 14 from helium 4.

the periodic table, including Carbon, Nitrogen, Oxygen, and other elements of life, are formed from hydrogen in the deep interiors of stars. This is considered to be one of the most monumental discoveries of the century!

Firm beliefs about cosmic dust were already set by the work of Dutch astronomers of the 1940's, and almost set in stone. Every astronomer from that time on believed that the bulk of interstellar dust had to be formed *in situ* in the gigantic diffuse gas clouds of space which had densities amounting to less than a single atom in a cubic metre. The composition of this cosmic dust was also thought to be beyond any dispute — composed of water ice with smaller admixtures of other inorganic ices — very similar to the ice particles that populate the cumulous clouds of Earth's atmosphere. Of course, it is true that oxygen is more abundant than carbon everywhere in space, so most of the carbon of interstellar space one might argue would tend to be tied up in the strongly bound molecule carbon monoxide (CO) and not available for condensing as dust. This would be the case if thermodynamic equilibrium prevailed, and it goes without saying that non-thermodynamic processes — of which biology is one —

were not even remotely considered at the time. This was therefore an apparently plausible argument at the time against the acceptance of a carbonaceous composition of interstellar dust — certainly a biological origin was not even considered at the time.

Stars emit visible light as well as radiation that is invisible to the eye — from X rays and ultraviolet to infrared, microwave and even radio wavelengths. The distribution of the energy radiated at these various wavelengths depends on the temperature of the radiating objects. The sun, which is at temperature of about 6000°C at its surface, emits most of its radiation in the visible part of the spectrum. This is the circumstance that gives the chance for life to exist on Earth. The main absorption wavelength of chlorophyll, the green colouring substance in plants, is close to the peak emission wavelength of the sun. The sun's energy is thus efficiently taken up by plants to convert water and carbon dioxide into carbohydrates, and this non-thermodynamic process is at the base of the food chain for all life on Earth. Other stars, if they are cooler, emit most of their energy in the far-red or infrared wavelength range; if they are much hotter they would emit mainly in the ultraviolet. Superposed on this continuum of radiation emitted by stars is a host of very sharp and narrow absorption or emission lines, giving astronomers all the information necessary to deduce the types of chemical elements of which stars are made.

Starlight travels vast distances through space, hundreds of light years, to reach us. (One light year is the distance traversed by light travelling at a speed of 300 million metres per second.) On the way to Earth this light must cross one or more interstellar dust clouds (Fig. 5.1). If the clouds happen to be too thick in dust they may effectively block out *all* the visible light from the stars. This is indeed the reason for the dark appearance of clouds of various shapes — elephant's trunks, horse's heads or eagle's nests — seen against the background of stars in the Milky Way. For clouds that are not so dense, the light from stars traversing through them is partially absorbed and scattered by dust particles in ways that leave tell-tale signs and signatures of their properties on the light that is received by an observer on Earth. The dim light of a street lamp seen through a fog has imprint on it, tell-tale signs that reveal the chemical properties and sizes of the intervening dust particles and molecules in the air. In the same way the dust in interstellar space leaves its own imprint on the transmitted light from stars, and this process of deciphering this imprint is what eventually led to the stark conclusion (we shall come to later) that bacteria and viruses in various states of decay and degradation exist throughout space.

Unravelling the nature of cosmic dust in this way requires (1) astronomical observations of stars over a wide range of wavelengths, (2) modelling the possible or plausible compositions of the dust, and (3) performing calculations of the scattering and absorption properties of dust in order to discriminate between various models. This is precisely the type of research that was to occupy a large part of the professional life of one of the present authors (CW) over the period 1962–1982, and led eventually to the theory of cosmic panspermia and a new theory of the origins of life.

The dark clouds of interstellar space had, however, to be set in the context of what else there existed in the Universe. The Milky Way itself, against which these clouds can be seen, our Galaxy, is a few hundred thousand light year-wide collection of hundreds of billions of stars, each one more or less similar to the sun. And our galaxy is one of many billions of similar galaxies that populate the observable Universe that is unimaginably vast. Interstellar dust clouds are by no means restricted to our own galaxy — our Milky Way system. External galaxies often show conspicuous dust lanes. We know that interstellar clouds in our own galaxy as well as in external galaxies are stellar nurseries, the birthplace of new stars and planets. Therefore, they must clearly have a very special importance in the scheme of things — an importance that was not fully recognised in the 1960's.

An interstellar cloud is on the average about 10 light years across in size, and the typical separation between neighbouring clouds is about 300 light years. There is a fairly wide spread of sizes of these clouds and in their structure: some are more compact and uniform in their disposition, whilst others are extended and irregular. The more extended clouds appear as giant complexes, showing a great deal of fine-scale structure as cloudlets and filaments. These are the so-called "giant molecular clouds" of which the molecular complex in the Orion nebula is an example.

Interstellar clouds may contain anywhere from ten to many millions of individual atoms per cubic centimetre. Even the higher values in this density range are considerably lower than the densities that can be attained in laboratory vacuum systems. So, it should be remembered that all our intuitive ideas of how gases behave under normal conditions may prove wide off the mark when it comes to understanding what happens under the extremely rarefied conditions of space.

Hydrogen makes up the overwhelming bulk of material in interstellar clouds and occurs in one of three forms: neutral atomic hydrogen (intact atoms with no electrons lost), ionized hydrogen (atoms stripped of their outer electrons), and molecular hydrogen (atoms paired in molecular form as H_2). Molecular hydrogen was first detected using ultraviolet spectroscopy in the late 1960's after CW's researches into interstellar matter had begun, although its existence was predicted by Fred Hoyle in the 1950's and indeed exploited in the science fiction novel *The Black Cloud*. Hydrogen molecules are to be found mostly in the denser interstellar clouds that are able to screen off the ultraviolet starlight that would otherwise destroy them. A large fraction of all the hydrogen in the galaxy is found to be in molecular form, H_2, and the total mass of the molecular hydrogen is billions of times the mass of the sun.

But what else is there in interstellar clouds besides hydrogen? Information derived from several sources, including studies of the spectra of the sun and stars, and the direct examination of meteorites (rocks of extraterrestrial origin) all have a bearing on unravelling the overall composition of interstellar material. Next to hydrogen, in order of abundances, comes the element helium, which accounts for close to a quarter of the total mass of interstellar matter. But from our point of view this element is inert, chemically unreactive and therefore uninteresting. Then comes the group of chemical elements carbon, nitrogen and oxygen that together make up several percent of the mass of all the interstellar matter. It is these elements that are of course crucial for life. Indeed, life depends on the function of the unique range of properties of the carbon atom including its high levels of chemical reactivity and its ability to combine into many millions of interesting carbon-based compounds. Next in line are the elements magnesium, silicon, iron and aluminium, which again account for a percent or so of the total interstellar mass. Then a group including calcium, sodium, potassium, phosphorus is followed by a host of other less abundant atomic species. All these chemical elements are synthesised from hydrogen in the deep interiors of stars in the manner worked out by Fred Hoyle and his colleagues in the 1950's. The synthesised elements are injected into interstellar space through a variety of processes, including mass flows from the surfaces of stars. In the case of the most massive stars, the end product of their evolution is a supernova (an exploding star), and it is through supernova explosions that life-forming chemical elements are mostly injected into the interstellar clouds and distributed throughout the universe.

The discovery of interstellar molecules (assemblages of atoms) by methods of radioastronomy, millimetre-wave astronomy and infrared astronomy got properly under way in the 1970's and 1980's. Next to molecular hydrogen the second most abundant and widespread molecule in space turns out to be carbon monoxide. Then most significantly for our story is the vast array of complex organic molecules present in huge quantity, and these discoveries as they progressed through the 1980's and 1990's provided the basis of some of the evidence that supported the concept of panspermia that we shall review later. In the denser interstellar clouds, particularly in clouds associated with new-born stars, vast amounts of water in gaseous form are found. Water is an important molecule for life, and its close association, along with the organics, with newly formed stars and planetary systems would have a vital relevance to our story.

The spatial distribution of interstellar molecules in the galaxy shows wide variations depending on physical conditions such as ambient temperature and density as well as the proximity of clouds to hot stars. As a rule, denser and cooler clouds contain the larger and more complex molecules, whereas lower density clouds and those nearer to hot stars have simpler molecular structures. A region that is particularly rich in organic molecules (molecules that could be connected with life) is the complex dust clouds in the constellation of Sagittarius, located near the centre of the galaxy. It is in this region that the first detection of an interstellar amino acid, glycine, (a component of proteins) was reported, as was the molecule of vinegar and a sugar glycolaldehyde.

An important class of organic molecule that is found to be present in vast quantity in space are the so-called polyaromatic hydrocarbons (or PAH's). These molecules (just as CO) are by-products of the combustion of fossil fuels as occurs for instance in automobile engines. They are essentially combustion/degradation products of living material and are the substances that are largely responsible for the suffocating smog that pollutes our larger industrial cities. Even in interstellar space it is plausible to suggest (in our view) that such molecules are most likely to have a direct biological connotation, possibly representing the break-up or degradation products of biological material.

Comparatively high densities of organic molecules tend to be associated with regions of the galaxy where new stars (and presumably comets and planets) are forming at a rapid rate. The Orion nebula is a spectacular example of such a region where young stars are evident in large numbers, some even with discs of newly formed planetary material seen around them. Large quantities of organic molecules are associated with the denser parts of the Orion cloud complex, and it would be tempting to link the formation of such molecules with the formation of stars, planetary systems, and perhaps life itself.

In addition to atoms and molecules, interstellar clouds contain an all-pervasive and enigmatic component in the form of dust that will provide a vital clue in support of panspermia. Astronomers had struggled to understand the precise nature of this cosmic dust and to discover the circumstances under which such particles are formed. In the autumn of 1961 when one of us (CW) started reading on these matters in earnest he found that it was almost an article of faith amongst astronomers that interstellar dust grains were comprised of dirty ice material — frozen water with perhaps a sprinkling of other ices — ammonia-ice, methane-ice and a trace quantity of metals. Furthermore, the firmly-held belief was that these particles had to condense, more or less continuously from the gaseous atoms and molecules that were present in the interstellar clouds. This was not justified by the well-established science (science of nucleation) that had existed at the time. Formation of solid particles from a tenuous gas was a two stage process. It required (a) nucleation (seeding) and then (b) condensation of gaseous molecules upon "seeds" or "condensation nuclei".

It was soon discovered that the insuperable difficulty of forming condensation nuclei for dust to form in the exceedingly tenuous gas clouds of interstellar space made the then fashionable ice grain theory essentially untenable. It is at this point that the story of interstellar dust took a radically different turn. The question arose: what if the cosmic dust was not made of water ice (as was the general belief at the time) but was made of carbon? At this stage the suggestion was that the formation of carbon dust occurred at much higher temperatures, perhaps, for instance in the outer atmospheres or envelopes of some cool carbon-rich stars? Alternatively, biologic dust could be generated in vast quantity in cosmic "breweries" in comets. But this latter possibility did not take shape until decades later.

But what did the astronomical observations in the 1960's and 1970's really tell us about the properties of interstellar dust? What are their precise optical characteristics as judged from an astronomical perspective? We have already noted that dust shows up as conspicuous patches of obscuration against the background of distant stars. But several more precise quantitative statements about the nature of the dust were already possible by the early 1960s.

The earliest attempts to obtain measurements of interstellar dimming — or extinction, as it is called — of starlight were made as far back as the 1930's. In the 1960's the most extensive modern data on interstellar extinction came from the meticulous observations of the Indian Astronomer Kashi Nandy at the Royal Observatory in Edinburgh. It was found that at a single wavelength close to 4500 Å (in the blue region of the spectrum) the dimming of starlight amounted to a reduction of intensity by a factor of about two for every 3000 light years of traversal through interstellar space. From this one piece of information alone it was easy to infer that interstellar dimming could only be reasonably attributed to solid particles that have dimensions comparable to the wavelength of visible light, that is a few thousand Angstroms ($1\text{ Å} = 10^{-8}$ cm) — matching exactly the sizes of bacteria on the Earth.

With the advent of new techniques in observational astronomy it became possible to measure with an increasing accuracy how interstellar dimming caused by cosmic dust varies with the wavelength of light. This relationship between interstellar dimming or extinction and the wavelength of light — what astronomers call the extinction curve — provides an important item of information to tell us about the properties of interstellar dust grains. In 1961 the extinction curve was known only over a very limited range of wavelengths — from about 9000 Å in the red end of the spectrum to about 3300 Å in the near ultraviolet.

Over much of the visible wavelength range it was known that the opacity (light absorbing properties) of interstellar dust was proportional to the inverse of the wavelength — in other words, when the wavelength is doubled, the opacity was approximately halved. And most remarkably, it turns out that precisely the same type of relationship, the same extinction curve, was found to hold over wide areas of the sky. This means that we must have dust with almost identical sizes and properties throughout large volumes of galactic space.

The limited range of wavelengths for which the so-called "extinction" data was available in 1961 allowed for a wide range of possible dust models, including ice (of the type proposed earlier by the Dutch astronomer H.C. van de Hulst) and iron (of the kind suggested even earlier by the Swedish astronomer, C. Schalen). For each of these models, however, the sizes of the dust particles had to be fixed within a very narrow range. To match the astronomical extinction data to any particular model one needed to calculate the scattering and absorption properties of particles of various radii using Electromagnetic Theory as developed by Maxwell a century or more ago. Such calculations that had already been made for iron grains and ice particles established these models as being *possible* candidates for interstellar dust, with ice particles having the edge over iron in certain crucial respects. Particularly so because it could be argued that there was insufficient iron in the galaxy to make up the required mass in the form of the interstellar dust.

In 1961–1962, in the days before personal computers, the calculation of the optical properties of any new grain model was a major computational undertaking, using formulae derived by the German physicist Gustav Mie in 1908. The task at this time was to turn these so-called Mie formulae into computer programmes for use on high speed electronic computers such as were available for example at the University of Cambridge. This is what one of us (CW) set out to do in the winter of 1961 for a preliminary exploration of the new carbon dust models.

This computation for carbon (graphite) dust in the 1960's showed that as long as the diameter of carbon particles were less than about a tenth of a micrometre, their predicted extinction behaviour was almost indistinguishable from the interstellar extinction observations as they were then known. It was also possible to calculate how much in the form of carbon grains we required, and the answer turned out to be one to two percent of the total mass of the interstellar clouds. This was consistent with the availability of carbon in interstellar space.

The next question to be asked was: where are the carbon grains produced? Searching for suitable cosmic locations, it seemed natural to turn to cool giant stars that are carbon rich. These are the carbon stars that are classified into two types — the N and R series, where the N series stars are older stars that show element (Ba, Sr) enhancement over the R stars. Optical spectra of these stars show the C_2 Swan bands and also spectral features showing C_3, CN, CH, SiC_2.

The N-stars have surface temperatures that varied cyclically between 1800 and 2500 K over a period of about a year, and they are known to have an excess of atmospheric carbon over oxygen. Thus, although oxygen would link up with C to form the strongly bound molecule CO (carbon monoxide), there would be an excess of C that is available to condense into solid particles when the temperature fell below some critical value. Again, a computer programme had to be used to determine the physical state of the excess carbon as the temperature varied between 1800 and 2500 K in the star's pulsation cycle. Was the carbon in a gaseous or solid state? It was quickly shown that solid carbon particles were indeed able to nucleate and grow out of the carbon gas in the stellar atmosphere as soon as the temperature fell towards 2000 K. It was further found that particles of radii of a few hundred Angstroms would grow and be expelled into interstellar space, the expulsion being caused by the pressure exerted by starlight from the parent star. There was strong evidence from astronomy that pointed to the existence of dust around such carbon-rich stars. The variable and highly luminous carbon star R Corona Borealis is a spectacular example. Here we see direct evidence of a star erratically puffing out clouds of carbon soot into the interstellar medium.

This work which formed the basis of the PhD dissertation by one of us (CW) was initially published as a joint paper with Fred Hoyle entitled "On graphite particles as interstellar grains" in the *Monthly Notices of the Royal Astronomical Society*. The publication represented the first step in the direction of cosmic biology although this was not recognised as such at the time. Soon afterwards CW made a prediction of the properties of interstellar grains based on new measurements of the optical constants of graphite. The prediction was that if the extinction curve of starlight (the dimming properties of dust) was extended into the ultraviolet, a strong absorption feature centred at a particular wavelength 2200 Å would be seen. When the first observations of stars from above the atmosphere were made, this predicted feature was discovered, and it later served as a spectroscopic beacon for identifying biology even in the most distant cosmic locations. It became clear that a major paradigm change was around the corner — a shift from volatile ice grains to cosmic dust grains that had to be based largely on the element carbon.

From Dust to Organic Molecules

The next steps in the progress towards unravelling the precise nature of cosmic dust required observations of stars that extended outside the visible range of wavelengths. New techniques in observational astronomy were now making it possible to study the behaviour of interstellar dust at longer wavelengths beyond (the infrared) and shorter wavelengths beyond blue (the ultraviolet). By 1965 the absence of an infrared absorption band in the spectra of stars near the wavelength 3.1 μm, the diagnostic of water ice, led to the conclusion that ice particles if they exist at all can make at most only a very minor contribution to the interstellar dust.

The observations of stars using telescopes on the ground were affected by absorption of light as it traverses the Earth's atmosphere. Essentially all the ultraviolet light of stars was filtered out in ground-based observations. With the dawn of the Space Age in the mid 1960's, astronomical observations were possible from above the atmosphere using equipment carried on rockets and satellites. Such observations of stars in the ultraviolet revealed a conspicuous feature in the form of

Fig. 5.4. CW and Fred Hoyle taken in 1979.

a broad peak of absorption centred on a precisely defined wavelength 2175 Å. This was exactly the property calculated for graphitic carbon particles of a particular size — of radius 0.02 µm. The old ice grain model of cosmic dust could not reproduce such an ultraviolet feature and must therefore be deemed to be inconsistent with observations and abandoned. A new generation of telescopes and instruments carried aboard satellites (for example, NASA's Orbiting Astronomical Observatory 2 (OAO2)) was soon to come into full operation confirming this conclusion beyond a shadow of doubt; so the die was cast to follow the trail that led eventually to cosmic life.

A striking result that emerged from the new studies was the invariance of the 2175 Å absorption hump from star to star, thus declaring the universality of some chemical component of the cosmic dust. This invariance of the ultraviolet feature, according to our original carbon grain models, demanded the presence of graphite particles in the form of spheres with radii fixed almost at a definite value, 0.02 µm. Even as early as 1969 astronomers were beginning to

Fig. 5.5. Hubble telescope image of the Orion nebula.

feel uneasy about the artificiality of such a constraint, so the lead that one of us (CW) next followed was to identify this ultraviolet feature with a spectroscopic property of the material in the dust, and that looking more and more like complex organic molecules in space.

Figure 5.5 shows the Orion Nebula which contains giant clouds choc-a-bloc with organic molecules. Here is an active site of star-births, the youngest stars being younger than a few million years, and including many nascent planetary systems (protoplanetary nebulae). This veritable stellar and planetary nursery is considered by many to be a region where pre-biotic chemistry occurs on a grand cosmic scale. It was argued by CW and others as an alternative that this may more plausibly represent a graveyard of life — polyaromatic hydrocarbons and other organic molecules discovered here arising from the destruction and degradation of bacterial life.

By the mid 1970's it was becoming amply clear that the best agreement for a range of astronomical spectra embracing a wide wavelength interval turned out to be a dust composition that is indistinguishable from freeze-dried bacteria and the best overall agreement over the entire profile of interstellar extinction was a mixture of desiccated bacteria, nanobacteria, including biologically derived aromatic molecules as seen in Fig. 5.6. The same feature characteristic of biology was also discovered widely outside our own galaxy as shown in Fig. 5.7.

Although many astronomers continued to look for abiotic (non-biological) models to explain the data such as in Figs. 5.6 and 5.7, biology emerged at this time to be by far the simplest self-consistent model. In particular, the claim that the strong peak of interstellar extinction at 2175Å can be explained by aromatics (PAH's) uncon-nected with life could be seriously flawed and could be easily disposed of. Aromatic molecules resulting from the decay, degradation or combustion of biomaterial may be similar to soot or anthracite. Facts relating to the widespread distribution of microbial life and their degradation products continued to come thick and fast.

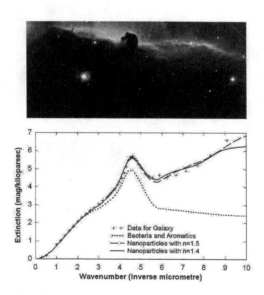

Fig. 5.6. (Top) Horsehead Nebula in Orion showing clouds of cosmic dust.
(Bottom) Agreement between interstellar extinction (dimming data) (plus signs) and biological models of cosmic dust. The 2175Å hump in the extinction is caused by biological aromatic molecules.

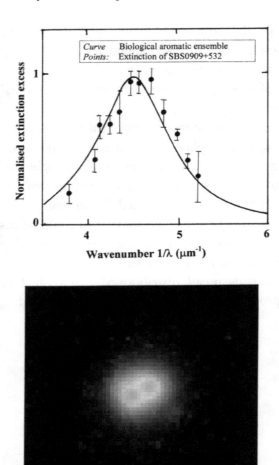

Fig. 5.7. Agreement between the 2175Å absorption of biomolecules (life) and the dust dimming profile of the galaxy SBS0909+532 at a distance of nearly 8 billion light years.

The paradigm in the 1960's that cosmic dust was largely comprised of water-ice was quickly being overturned in the 1970's with the advent of infrared observations showing absorptions due to CH, OH, C–O–C linkages consistent with organic polymers. The best agreement

for a range of astronomical spectra embracing a wide wavelength interval turned out to be material that is indistinguishable from freeze-dried bacteria and the best overall agreement over the entire profile of interstellar extinction was a mixture of desiccated bacteria, nanobacteria, including biologically derived aromatic molecules.

The birth of the new discipline of infrared astronomy now came into play. The first discovery of infrared astronomy that became directly relevant to our story came from the work of John E. Gaustad and his colleagues who confirmed that the infrared spectra of highly reddened (dimmed) stars showed no evidence whatsoever of water ice. This was a victory for the organic dust theory and a further disproof for the old ice grain theory. Next there followed a spate of discoveries all showing the presence of a new spectral feature of dust over the infrared waveband 8–12 µm. The feature was observed in emission in a wide variety of astronomical objects and it was immediately interpreted as evidence that the cosmic dust was made of a mixture of silicates — combinations of magnesium, silicon and oxygen as they occur in the rocks of the Earth. Inorganic material, now in the form of silicates, soon became the fashion in astronomy — a fashion that one of us began to question. Was the newly discovered 8–12 µm absorption/emission in cosmic dust really due to silicates? Is it not possible that some organic materials were responsible for absorptions over the same waveband?

On closer investigation of the silicate model of dust it soon became apparent how poorly the newly observed astronomical feature over the 8–12 µm band actually fitted the behaviour of any known silicate or mixtures of silicates. Because mineral silicates (e.g. rocks) do indeed have absorptions spanning the 8–13 µm waveband, there was a crude match to be seen. But the identification of silicates in vast quantity in interstellar space was by no means compelling at this point. Other chemical systems, including some that involved carbon, could be far stronger candidates. Whilst one could not dispute that some quantity of silicate dust might exist in space, how would this compare with contributions from the far more cosmically abundant element carbon? This was the question that was explored for over a decade before arriving at the conclusion that complex organic material, and even biological material, might be involved in producing the infrared signatures that were wrongly being attributed to mineral silicates.

The search for a possible carbon-based interpretation of the 8–13 μm interstellar feature began as early as 1969. What if the carbonaceous component of the dust was not simply graphite or solid carbon as was proposed in 1962 but made of organic materials, organic polymers in fact — even biopoymers? Perhaps carbon atoms in interstellar space might be combined with hydrogen and oxygen to form an extraordinarily vast variety of organic chemicals. In terms of the basic chemical elements at least there would be more than enough mass to explain the properties of interstellar dust.

At this time the molecule formaldehyde H_2CO had been discovered to exist ubiquitously as a gaseous molecule in interstellar clouds. It was present in dense molecular clouds as well as in the less dense interstellar medium. What if such molecules started to condense and polymerise on the surfaces of pre-existing graphite or silicate grains expelled from stars? It turned out that such polymers were dielectric (that is to say, non-absorbing) in the visual waveband as the astronomical observations demanded. And most strikingly polyformaldehyde (polyoxymethylene) had absorption bands over the 8–12 and 16–22 μm wavebands, with the former absorption fitting the astronomical data far better than any known silicate. This was a breakthrough moment and within a couple of weeks the paper entitled "Formaldehyde Polymers in Interstellar Grains" was written and submitted to *Nature*.

When it was published in *Nature*, it made a major splash in the science news columns of the broadsheets. This was the first-ever suggestion of the widespread occurrence of organic polymers in the galaxy. It marked the beginning of the cosmic life theory that was developed throughout the 1970's and 1980's. With the accumulation of more infrared data from astronomical sources it became clear that there was unequivocal evidence of organic polymers existing on a vast scale throughout the galaxy. Models involving co-polymers — mixed chains of formaldehyde and other molecules as well as polymer mixtures resembling tars and biopolymers — were all considered, leading inexorably to one direction — life. What if the interstellar grains that CW had begun investigating in 1960 were indeed connected with biology, with life itself? This question itself, with all its profound implications, was indeed a brutal assault on conventional scientific thought.

A crucial breakthrough came when the Cardiff based team led by CW collaborated with D.T. Wickramasinghe, (CW's brother), Professor of Mathematics at the Australian National University in Canberra and an astronomer who had access to the 3.9 m Anglo-Australian Telescope, in Siding Springs. This telescope at the time happened to be equipped with just the right instruments to look for an infrared signature of interstellar bacteria in deep space.

In February and April 1980, Dayal, collaborating with D.A. Allen, obtained the first spectra of a source known as GC-IRS7 which showed a broad absorption feature centred at about 3.4 µm. When Dayal's published spectrum was examined we found that it agreed in a general way with the spectrum of a bacterium that we had found in the published literature. But at this time neither the wavelength definition of the astronomical spectrum nor the laboratory bacterial spectrum was good enough to make a strong case for interstellar bacteria. However, even from this early observation we were able to check that the overwhelming bulk of interstellar dust must have a complex organic composition.

It was precisely at this moment that Shirwan Al-Mufti with laboratory experience became a research student at Cardiff. Here was the chance to get the required laboratory work done. Al-Mufti's experiments involved desiccating bacteria, such as the common organism *E. Coli*, in an oven in the absence of air and measuring, as accurately as possible, the manner in which light at infrared wavelengths is absorbed. The normal technique for doing such a measurement involved embedding the bacteria in discs of compressed

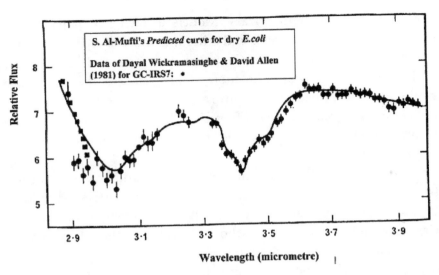

Fig. 5.8. S. Al-Muftis' Predicted Curve for Dry *E. coli*, later to be matched by observation as we shall see.

potassium bromide and shining a beam of infrared light through them. The standard techniques had to be adapted only slightly. There was the need to match the interstellar environment which involved desiccation, and the spectrometer that was used had to be calibrated with greater care than a chemist would normally exercise. When all this was done, it turned out that a highly specific absorption pattern emerged over the 3.3 to 3.6 µm wavelength region, and this pattern was found to be independent of the type of microorganism that was looked at. Thus, whether we looked at *E. Coli* or dried yeast cells it did not matter. This precise invariance came as a great surprise. The newly discovered invariant spectral signature was a property of the detailed way in which carbon and hydrogen linkages were distributed in biological systems and could be looked for in astronomy.

The observations that were to mark a crucial turning point in this entire story were carried out *after* an experimental prediction was made following Al-Mufti's experiments. The new observations were of a superior quality because a new generation of spectrometers had become available. Dayal Wickramasinghe sent CW his raw data by fax to compare with the new laboratory spectra which had been obtained just months earlier by Al-Mufti in March and April of the same year. After an hour or so of straightforward calculations it was possible to overlay the astronomical spectrum over the detailed predictions of the bacterial model and find a staggeringly close fit. This was perhaps the best possible confirmation of the bacteria model of cosmic dust, particularly because the experimental data in the comparison was obtained *before* the final astronomical observations became available. A precise agreement between a set of data points and a predicted curve is normally regarded as a consistency check of the model on which the curve is based — QED. But in this case, since the model of bacterial grains runs counter to a major paradigm in science, the situation was otherwise.

Even as early as 1962, the presence of aromatic molecules in space might have been inferred from the so-called diffuse interstellar absorption bands. It has been known for over half a century that some 20 or more diffuse absorption bands appear in the spectra of stars, the strongest being centred on the wavelength 4430 Å. Despite a sustained effort by scientists over many years no satisfactory inorganic explanation for these bands has emerged. A possible solution was offered in the 1960's by the chemist F.M. Johnson who showed that a molecule related to chlorophyll-magnesium tetra-benzo porphyrin has all the required spectral properties to explain the astronomical data. Chlorophyll is, of course, an all-important component of terrestrial biology — it is the green colouring substance of plants, the molecule responsible for photosynthesis, the process that lies at the very base of our entire ecosystem on Earth. So its abundance in deep space has a direct relevance to the theory of cosmic biology.

Very recently scientists have unearthed yet another property of biological pigments such as chlorophylls, a property that clearly shows up in astronomy. Many biological pigments are known to fluoresce; in the fashion of pigments as such exist in glow worms. They can absorb blue and ultraviolet radiation and fluoresce over a characteristic band in the red part of the spectrum. For some years now astronomers have been detecting a broad red emission feature of interstellar dust over the waveband 6000–7500 Å. Chloroplasts contain chlorophyll when they are cooled to temperatures appropriate to interstellar space fluoresce precisely over the same waveband. Cosmic biology announces itself once again in the form of fluorescing pigments, similar to those found in many biological systems.

6

Dawn of Modern Astrobiology

By the start of the 1980's one of us (CW) and Fred Hoyle were firmly committed to the view that an immensely powerful cosmic biology somehow came to be overlaid on Earth from the external universe some 4 billion years ago. In 1982 at a public lecture delivered in Cardiff under the auspices of the Royal Astronomical Society Fred Hoyle concluded thus:

> Microbiology may be said to have had its beginnings in the nineteen-forties. A new world of the most astonishing complexity began then to be revealed. In retrospect I find it remarkable that microbiologists did not at once recognise that the world into which they had penetrated had of necessity to be of cosmic order. I suspect that the cosmic quality of microbiology will seem as obvious to future generations as the Sun being the centre of the solar system seems obvious to the present generation.

At the same time Fred Hoyle and CW wrote that the birth of a new scientific discipline combining astronomy and biology was imminent and he suggested the name Astrobiology — a fact that is ignored by modern commentators who wish to stake their claim for the new discipline.

The Earth is just one of many billions of habitable planets in the galaxy that are now known to exist, following observations using the Kepler space telescope. Our planet served as an assembly station for the cosmic genes that carry the blueprint of all conceivable life forms in the cosmos. The agency responsible for the transfer of such genes

71

are the comets in which warm radioactively heated domains serve as repositories for cosmic microorganisms — bacteria and viruses.

An individual comet can range in size from a few to tens of kilometres, and a few hundred billion of such objects exist in our solar system mainly in a gigantic shell surrounding the entire solar system at a distance of about a light year from the sun. From time to time comets from this shell are deflected into orbits that bring them into the inner regions of the solar system, sometimes colliding with planets, but mostly to distribute material evaporating from the surface to form meteor streams that can be crossed by planets like Earth.

We suspected then (and know now) that cometary bodies contain long-lived radioactive heat producing elements such as Uranium and Thorium that can maintain radioactively heated liquid domains over many billions of years. Such liquid domains replete with organics and other nutrients serve, in our model as the repositories and amplifiers of cosmic life which exist in the form of archaea, bacteria and viruses. Such bacteria and viruses carry the genetic blueprint of all the permitted and permissible lifeforms in the universe. Besides the Earth other planetary bodies, within the solar system and elsewhere, must also be exposed to the process of contamination by cometary microbes.

We now know that the earliest evidence of microbial life in the form of bacterial residues exists in an outcrop of rocks in Western Australia dated 4.2 billion years ago. This was shortly after the formation of a rocky crust of the planet was complete, and the planet was in the midst of the Hadean epoch involving frequent collisions with comets. It is clear that impacting comets brought the first bacterial life to Earth, and the cometary impacts also led to the formation of oceans and an atmosphere around the planet. Life thereafter proceeds to evolve in fits and starts from bacteria, to metazoa, to plants and animals and eventually of course humans. This process occupied a timespan of over 4 billion years.

As we have seen, how life on Earth progressed from single-celled bacteria to multi-celled life and the enormous diversity of life we see

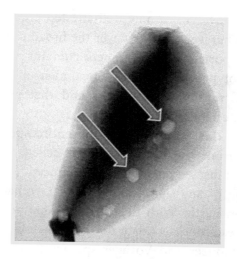

Fig. 6.1. Micron sized carbon spherules which are inferred to be relics of bacteria within zirconium crystal from the rocks in the Jack Hills outcrop in W. Australia.

today is a question that has occupied the minds of thinkers from the earliest times. Darwinian evolution is one component of the solution but by no means the whole. New innovations in the development of life, according to this point of view, are importation of viral genes from the cosmos involving a now well-attested process of "horizontal gene transfer". Wherever and whenever the broad range of the cosmic life system transported via the comets contains viruses defining a form of life (genotype) that matches a local niche of a recipient planet, horizontal gene transfer would occur and that particular lifeform could succeed in establishing itself.

According to this point of view the entire spectrum of life on Earth, ranging from the humblest single-celled life forms to the higher animals must essentially be introduced from the external cosmos. Nothing of any significance could have happened within a closed system on Earth. With this in mind, one of us together with Hoyle began to examine new data on the planets of our solar system obtained from Pioneer and Voyager spacecraft for tell-tale signs of microbial life. An indispensable condition for life was however the need to have access to liquid water.

Subject to this constraint it was found that there were tentative signatures of bacterial life in the planets Venus, Jupiter and Saturn already by the early 1980's. Since Venus is exceedingly hot at ground level (about 450°C) it would be impossible for life to exist at the surface. Venus, however, has an extensive cloud cover and it is within these clouds that life may have taken root. Water is present in small quantities and in the higher atmosphere the temperature is low enough for water droplets to form. Moreover, the clouds of Venus are in convective motion in the upper atmosphere, which ranges in height between 70 km to 45 km, with a corresponding temperature range of 75°C at the top to −25°C at the bottom. It was argued that the survival of bacteria over the range of conditions in the upper atmosphere was possible, and that repeated variations of temperature in a circulating cloud system would tend to favour bacteria capable of forming sturdy spores. A strong case was given for an atmospheric circulation of bacteria on Venus between the dry lower clouds and the wetter upper clouds where replication might take place. It was discovered that Pioneer spacecraft data, including the presence of a rainbow in the upper clouds, could be interpreted as implying the presence of scattering particles that had precisely the properties appropriate to bacteria and bacterial spores.

These ideas have subsequently come into vogue. In 2002 Dirk Schulze-Maluch and Louis Irwin looked at data on Venus from the Russian Venera space missions and the US Pioneer Venus and Magellan probes. They discovered trademark signs of microbial life from studies of the chemical composition of Venus's atmosphere 30 miles above the surface. They expected to find high levels of carbon monoxide produced by sunlight but instead found hydrogen sulphide and sulphur dioxide, and carbonyl sulphide, a combination of gases normally not found together unless living organisms produce them. They conclude that microbes could be living in clouds 30 miles up in the Venusian atmosphere, exactly in the manner we discussed 25 years earlier (*New Scientist,* 26 September 2002).

Hoyle and CW had also argued for localised bacterial populations in Jupiter's atmosphere that might even have a controlling effect on its meteorology, including the persistence of the Great Red Spot. A kilometre-sized cometary object hitting Jupiter at high speed will be disintegrated into hot gas that would form a diffuse patch similar to the Great Red Spot. Such a region of the atmosphere would be rich in the inorganic nutrients needed for the replication of microorganisms. A large bacterial population could then be built up in this area and the possibility arises for a feedback interaction to be set up between the properties of the local bacterial population and the global meteorology of Jupiter as a whole. Additionally, a case was presented for bacterial grains trapped in the rings of Jupiter, Saturn and Uranus, rings that were discovered in 1979 by the Voyager missions.

It was also argued at this time that the presence of methane and other organic compounds in any quantity on solar system bodies as being an indication of life. Essentially all the organics on the Earth today are either directly or indirectly due to biology. So it is likely to be for organic matter that is found in substantial quantity elsewhere in the Universe. The outer planets, which are known to contain methane and other complex organic molecules, in their atmospheres, must be teeming with microorganisms according to our point of view. All these speculations were discussed at length in *Space Travellers: The Bringers of Life* (University College, Cardiff Press, 1981). It was in a closing section of this book that the inevitability of a merger between astronomy and biology was first published:

> The potential of bacteria to increase vastly in their number is enormous. It should occasion no surprise, therefore, that bacterial are widespread throughout astronomy. Rather it would be astonishing if biological evolution had been achieved on the Earth alone, without the explosive consequences of such a miracle ever being permitted to emerge into the Universe at large. How could the Universe ever be protected from such a devastating development? This indeed would be a double miracle, first of origin, and second of terrestrial containment.

Some biologists have probably found themselves in opposition to our arguments for the proprietary reason that it seemed as if an attempt were being made to swallow biology into astronomy. Their ranks may now be joined by those astronomers who see from these developments that a more realistic threat is to swallow up astronomy into biology. Clearly the solution is the merging of these ancient disciplines into a new discipline that can be called "astrobiology." The utter improbability of life's origin in a primordial soup generated in the primitive oceans is what drove one of us (CW) and Fred Hoyle irrevocably and implacably towards astrobiology and panspermia.

A functioning of a living system depends on many thousands of chemical reactions taking place within a membrane-bound cellular structure. Such reactions, grouped into metabolic pathways, have the ability to harness chemical energy from the surrounding medium in a series of very small steps: transporting small molecules into cells, building biopolymers of various sorts, and ultimately making copies of themselves while also possessing a capacity to evolve. Batteries of enzymes, composed of chains of amino acids in highly conserved arrangements, play a crucial role as catalysts precisely controlling the rates of chemical reactions. Without enzymes, there could be no life. It is the specificity and conserved information represented in the ordering of amino acids in enzymes that are absolutely crucial for life.

In the early decades of the twentieth century, the most important metabolic pathways in biology had already been unravelled, e.g., the carbon dioxide cycle in plants. However, even if we possessed a complete knowledge of all the metabolic pathways in biology, we would not come any closer to understanding the processes by which the simplest living system emerged. Many types of terrestrial venues for the origin of life have been explored over several decades — deep-sea thermal vents, primordial oceans amongst others — but all such sites of choice are grossly inadequate, in our view, to overcome the enormous hurdles of improbability that are involved.

In present-day biology, the information contained in enzymes — the arrangements of amino acids into folded chains — is absolutely crucial for life, and this information is transmitted by way of the coded ordering of nucleotides in DNA. In a hypothetical RNA world that may have predated the DNA — protein world in some origin of life models, RNA is posited to serve a dual role as both enzyme and genetic transmitter. If a few ribozymes are regarded as essential precursors to all life, one could attempt to make an estimate of the probability of the assembly of a simple ribozyme which is composed of some 300 bases. This probability turns out to be 1 in 4^{300}, which is equivalent to 1 in 10^{180} — which already defines an event that can hardly be supposed to happen even once in the entire 13.8-billion-year history of the universe. A similar calculation for the ordering of amino acids in a minimal set of bacterial enzymes gives an even more ridiculous probability ~1 in 10^{5000} with plausible assumptions.

On this basis it would appear impossible to avoid the conclusion that the emergence of the first *evolvable* cellular life form had of necessity to be a unique event in the cosmos. If this did indeed happen on Earth for the first time, it must be regarded as a near-miraculous event and one that could not be repeated elsewhere, let alone in any laboratory simulation on Earth. To overcome improbabilities on the scale that is involved here, it stands to common sense that one would gain immensely by going to the biggest system available — manifestly, the universe as a whole.

The argument that life as a cosmic phenomenon and panspermia must be rejected *a priori* because it merely transfers the problem of origin from Earth to another setting is by no means scientific. The question of whether life started *de novo* on Earth or was introduced from the wider universe is a scientific question that merits investigation and one that is open to testing and verification on many different levels. The invocation of Occam's razor to exclude such a discussion is merely an excuse for keeping scientific discussion within the strict bounds of orthodoxy — within the confines of Aristotelian philosophy incorporated in the fabric of Judeo-Christian theology in the early part of the first millennium. Occam's razor and other criticisms are being used to question any challenge to spontaneous generation strikingly reminiscent of the restrictions that stifled science in the Middle Ages.

By the early 1980's there was already an impressive body of evidence that pointed not only to the organic composition of cosmic dust but also to an origin of terrestrial life that had to be connected to the wider cosmos. It was therefore somewhat puzzling to understand the reluctance on the part of the scientific community to accept the facts. There were of course very much bigger issues at stake. If the whole of Darwinian evolution was to come under scrutiny, there would be a motive to turn away from even the simplest facts that pointed in such a direction. After all the victory of Darwinism over the narrow Judeo-Christian view of creation as exemplified in the famous Huxley–Wilberforce debate was a hard-won affair and the memory of the blood-letting must still linger in some form in our collective consciousness. It was a victory to be cherished at all costs, and "smaller" truths may need to be sacrificed in the interests of larger perceived goals. This, together with a deep cultural resistance to accept that life could have a deep connection with the external universe, continues to impede progress toward accepting the ideas that we shall describe in this book.

The next major project that Hoyle and CW undertook during the period 1980–1981 was an attempt to connect cosmic life, viruses and bacteria causing disease and coming from comets, with the process of the evolution of life on Earth. If life started on Earth some 4.2 billion years ago (current best estimate) with impacting comets bringing the first batch of cosmic microorganisms, how did such single celled microbes evolve and later diversify to yield the magnificent range of life forms we see today?

It is believed by traditional biologists that the full spectrum of life is the result of a primitive living system being sequentially copied billions upon billions of times. According to their theory, the accumulation of copying errors, sorted out by the processes of natural selection, the survival of the fittest, could account for both the rich variety of life and the steady upward progression of complexity and sophistication from a bacterium to man. This is perhaps a simple representation of Neo-Darwinism, but it encapsulates its essential features. Is this enough to explain all the available facts of biology? When Fred Hoyle and CW began to examine this question, their answer turned out to be an emphatic no.

In essence, the underlying argument was simple. Major evolutionary developments in biology require the generation of new high-grade information, and such information cannot arise from the closed-box evolutionary arguments that are currently in vogue. The same difficulty that exists for the origin of life from its organic building blocks applies also for every set of new genes needed for further evolutionary developments.

The alternative to the random assembly of life as a unique, perhaps unrepeatable event in a finite universe is assembly through the intervention of some form of cosmic intelligence. Such a concept would be rejected outright by many scientists, although there is no purely logical reason for such a rejection. With our present technical knowledge human biochemists and geneticists could now perform what even ten years ago would have been considered impossible feats of genetic manipulation. We could for instance splice bits of genes from one species into another, and even work out the possible outcomes of such splicings.

In 2010 a group of scientists led by Craig Venter in the USA have removed the entire genome of the bacterium *Mycoplasma genitalium* and replaced it with a synthetically assembled genome, producing essentially a modified bacterium which is able to replicate with the new artificial genome. It would not be too great a measure of extrapolation, or too great a license of imagination, to say that a cosmic intelligence that emerged naturally in the Universe may have designed and worked out all the logical consequences of our entire living system.

Hoyle and one of us (CW) also initiated a new form of panspermia that we have recently called "pathopanspermia". The arguments relating to cosmic evolution that we have discussed must connect also with the idea of disease-causing viruses coming from space. One might thus legitimately ask: if virus infections are bad for us why did the evolution of higher life not develop a strategy for excluding their ingress into our cells. Logically it seems easy enough for the greater information content of our cells to devise a way of blocking the effects of the much smaller information carried by a virus, and yet this has not happened in the long course of evolution. Could it be, one might wonder, whether this "invitation" to viruses was retained for the explicit purpose of future biological evolution? It is only with the dawn of the new millennium that an affirmative answer to this question was provided by data from the human genome project. A significant fraction of the human genome contains DNA derived from viruses — fragments of viral RNA sequences converted DNA form, and copied faithfully generation to generation. We might argue that this storehouse of extinct viral information could somehow carry the potential for future evolution. Moreover, there is further evidence from genome studies that our ancestral line was indeed attacked periodically with bacterial or viral infections that nearly culled the evolving line save for a small breeding group that survived and came through to modern times.

The human gut microbiota has been described as the most densely populated ecosystem on Earth, with advances in culture-independent gene sequencing techniques allowing us to estimate the total bacterial count of the human gut at around $\sim 10^{14}$ (100 trillion) individual cells. Automated gene sequencing has shown that the total number of genes associated with our microbiome massively exceeds the 22,000 or so protein coding genes in the human genome, while a gene catalogue of 3.3×10^6 non-redundant genes in the microbiome of the human gut has recently been published. Gene sequence mapping does not distinguish between viruses, plasmids and transposable genes, so it is possible that viruses make up the main component in the microbiome.

Gut bacteria divide into two broad classes: commensal or symbiotic organisms, and dysbiotic or pathogenic organisms. Included within these classes are many groups of extremophiles, including acidophiles, organism which have been shown to transfer viral particles (virions).

The potential role of the biome in health, immunity and disease is only now becoming recognized and health conditions that have defied understanding may well be connected with its functioning; patients with schizophrenia for example, have a significantly altered microbiome in their mouths and throats.

How did the human (and other) microbiomes arise in the first place? Is it the outcome of millions of years of co-evolution between evolved organisms (mammals, humans) and an ever-changing population of environmental bacteria and viruses? Or are the microbial/viral entities associated with microbiomes being continuously replenished from space, notably from comets?

From 2001 onwards evidence for an ongoing entry of biological entities from comets, consistent with Panspermia has grown to the point of being close to compelling. The case for Panspermia has been further strengthened by recent discoveries of exoplanets and the diminishing average distance between neighbouring habitable planets.

It is now no longer heretical to suggest that microbes, like the ones we have isolated from the stratosphere (discussed later), arrive from space and interact with human genome. The discovery of novel sequences that may be occasionally present in viruses could give a clue as to their space origin, thus establishing consistency with ongoing Panspermia.

In the balloon experiments that are being planned at the Institute for the Study of Panspermia and Astro-economics in-falling cometary material will be collected at regular intervals. Samples of the collected material will be used to look for evidence of microbiome-related genome components in stratospheric dust. Based on our 2001 sampling of the stratosphere at 41 km we estimate a daily input of biomaterial over the entire Earth of ~0.3 tonne. With a typical virion mass of 10^{-21} g we would thus have ~3×10^{26} virions per day. On this basis the total virion count in the oceans of ~10^{31} virions would be replaced on a timescale of 3×10^4 days ~100 yr, or three human generations. Over such a timescale space-derived microbiome-related virions can act as horizontal gene transfer (HGT) agents and play a crucial role in the evolution of their hosts; the evolution of hominids over millions of years, we suggest, has been driven by the accumulation of such space-derived virions into their genomes.

The recent COVID-19 pandemic has cemented in our minds the negative impact that viruses have on human welfare. Perhaps however, we should take a more appreciative view of these entities, since, as we have seen, we are certain that the evolution of humans, and all other life on Earth is the result of space-derived information delivered by these, and other microorganisms.

7

The Clues from Comets

Comets are relatively small celestial bodies that show up as spectacular objects in the night sky. Their conspicuous long tails stretch across great arcs in the night sky, and such events have had a long recorded history in Chinese, Indian and Egyptian annals. Comets were at once feared and revered in many ancient cultures, seeing them as harbingers of doom and bringers of pestilence and death. The true nature of individual comets was of course unknown to ancient cultures.

We now know that comets are in the main "icy" bodies but with a large component of rocky materials as well as organic molecules. Comets range in size from a few kilometres to hundreds of kilometres in the case of "giant comets". Recent studies have also shown that comets must have radioactively heated warm interior "lakes" that would have remained liquid for billions of years and thus served as the potential repositories of cosmic microbiology. Although an individual comet might be thought of as an insubstantial object there are a vast number of individual comets that have to be reckoned with. In the solar system alone we know that a giant shell of a hundred billion comets surrounds this system at a distance of a tenth of a light year from the sun. It is from this reservoir of comets that individual comets such as Comet Halley comes to be perturbed into orbits that take them to the inner parts of our planetary system, and indeed on occasion into our own vicinity.

Returning to our own story the next crucial step was indeed connected with comets and in particular with the return to perihelion (closest approach to the Sun) of Halley's comet in 1986. This was the first time that a comet was being studied by scientists since the beginning of the space age. From as early as 1982 a programme of international cooperation to investigate this comet came into full swing, the aim being to coordinate ground-based observations, satellite-based studies, and space-probe analysis on a worldwide basis. No less than five spacecraft dedicated to the study of Comet Halley were launched during 1985, the rendezvous dates being all clustered around early March 1986, about one month after the comet's closest approach to the sun.

The long-accepted theory of comets at this time was that they were dirty snowballs — the model of comets proposed by Fred Whipple. The expectation therefore was that the comet will appear from close quarters like a field of snow. On the night of March 13, 1986 scientists on the Giotto mission watched their television screens with nervous anticipation as cameras began to approach within 500 km of the comet's nucleus. The cameras were set up to photograph a bright snowfield scene on the nucleus consistent with the then fashionable Whipple dirty snowball model of comets. In the event the pictures transmitted worldwide on 13 March proved to be a disappointment, the cameras had their apertures shut down to a minimum and trained to find the brightest spot in the field. As a consequence, very little of any interest was immediately captured on camera — the scene was far too dark.

The much-publicised Giotto images of the nucleus of Comet Halley were obtained only after a great deal of image processing. The stark conclusion was the revelation of a cometary nucleus that was amazingly black. It was described at the time as being "blacker than the blackest coal...the lowest albedo of any surface in the solar system...." This was the vindication of a prediction that was a natural consequence of the organic/biologic model of comets. As we shall see more triumphs were soon to follow.

A few days after the Giotto rendezvous, infrared observations of the comet were made by Dayal Wickramasinghe and David Allen using the 40-metre Anglo-Australian Telescope. On March 31, 1986 they discovered a strong emission from heated organic dust over the 2 to 4 µm waveband. As noted earlier basic structures of organic molecules involving CH linkages absorb and emit radiation over the 3.3–3.5 µm infrared waveband, and for any assembly of complex organic molecules as in a bacterium, this absorption is broad and takes on a highly distinctive profile. The Comet Halley observations by Dayal and David Allen were found to be identical to the expected behaviour of desiccated bacteria heated to 320 K. Another victory for this model! Later analysis of data obtained from mass spectrometers aboard Giotto also showed a composition of the break-up fragments of dust as they struck the detector to be similar to bacterial degradation products (Fig. 7.1).

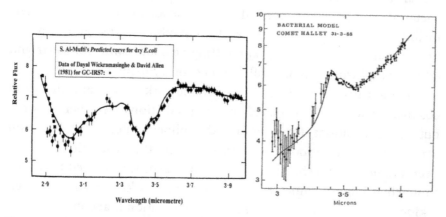

Fig. 7.1. Left panel: Comparison of the normalized infrared flux from GC-IRS7 with the laboratory spectrum of *E. coli* (already discussed).

The right panel in the above figure shows the emission by dust coma of Comet Halley observed by D.T. Wickramasinghe and D.A. Allen on March 31, 1986 compared with normalized fluxes for desiccated *E. coli* at an emission temperature of 320 K. The solid curve is for unirradiated bacteria; the dashed curve is for X-ray irradiated bacteria.

The Comet Halley observations of 1986 clearly disproved the fashionable Whipple's "dirty snowball" theory of comets although astronomers at the time were loath to admit this. The old theory still dies with variants of it postulating that there was more dirt (organic dirt) than snow! It could not be denied that water existed in comets in the form of ice, but great quantities of organic particles indistinguishable from bacteria are embedded within the ice. This conclusion was unavoidable unless one chose to ignore the new facts. More recent studies of other comets have yielded generally similar results. Most recently the European Space Agency's Rosetta Mission to comet 67P/C-G has provided the most detailed observations that satisfy all the consistency checks for biology and the theory of cometary panspermia. Figure 7.2 shows the close consistency between the surface properties of the comet and the spectrum of a desiccated bacterial sample. Other recent data from space exploration of comets include high abundances of the element phosphorus (indicative of life), a comet (Comet Lovejoy) emitting vast amounts of a sugar and ethyl alcohol which are natural products of fermentation.

By the early 1990's a conceptual framework for a grand theory of cosmic life was fully in place, and its predictions were being borne out in observations from several disciplines. Interstellar dust and cemetery dust were found to possess exactly the properties that had been predicted, and the oldest life on the Earth was pushed back to a time when intense cometary bombardment was known to have been taking place. We now have the strongest indication that comets seeded the planet with life some 4.2 billion years ago.

The discoveries of microbial life enduring the most extreme conditions were suggesting an alien context for all such properties, and opening possibilities of microbial habitats in a wide range of bodies in the solar system. A wrong theory does not come up repeatedly with such an amazing series of successes. Sooner or later a contradiction turns up and the theory has to be abandoned. This

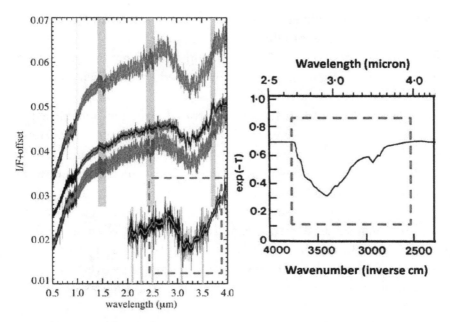

Fig. 7.2. The surface reflectivity spectra of comet 67P/C-G (left panel) compared with the transmittance curve measured for *E. coli* (right panel).

has not happened in our case. Why then is there such deep-rooted hostility to our ideas? The simplest answer was that such ideas went against the grain of an essentially geocentric scientific culture, and more importantly perhaps a long tradition of European supremacy in post-enlightenment science.

Despite all these impediments, CW together with Fred Hoyle and a small team of like-minded scientists continued to pursue these ideas doggedly in whichever direction that new data directed them. And new data did indeed come at a brisk pace. A discovery of a 3.28 μm emission feature in the infrared in the diffuse radiation emitted by the Galaxy confirmed that aromatic molecules of some kind were exceedingly common on a galactic scale. It was argued that the infrared emissions not just at 3.28 μm but over discrete set of wavelengths — 3.28, 6.2, 7.7, 8.6, 11.3 μm — must arise from the absorption of ultraviolet starlight by the same molecular system that degrades this energy into the infrared. Hoyle and one of us had shown much earlier that the 2175 Å extinction of starlight may be due to biological aromatic molecules, and it seemed natural then to connect the two phenomena.

This led to a unified theory of infrared emission and ultraviolet extinction by the same ensemble of aromatic molecules. A currently fashionable non-biological aromatic molecule was coronene ($C_{24}H_{12}$) and it was easy to demonstrate that this type of molecule was nowhere near as good as biological aromatics.

Our first major research undertaking was a collaboration with Indian scientists under the leadership of Jayant Narlikar and the Indian Space Research Organisation (Fig. 7.3). The aim was to detect bacteria entering the Earth by examining samples recovered from the high stratosphere. Initial laboratory work in Cardiff was carried out in the School of Biosciences with the involvement of David Lloyd and with the assistance of research student Melanie Harris. The first phase of this investigation was completed in July 2001 and we obtained unambiguous evidence for the presence of clumps of living cells in air samples from as high as 41 km, well above the local tropopause (16 km), above which no micron-sized aerosols from lower down would normally be expected to be transported. The detection was made using electron microscope images, and by application of a

Fig. 7.3. Chandra and his wife Priya, Jayant Narlikar and Milton Wainwright.

fluorescent dye known as carbocyanine that is only taken up by the membranes of living cells.

The variation with height of the density of such cells indicated strongly that the clumps of bacterial cells were falling from space. The input of such biological material was provisionally estimated to be between 1/3–1 tonne per day over the entire planet. If this amount of organic material was in the form of bacteria, the annual input of bacteria is a staggering 10^{21} in number amounting about a tenth of a tonne averaged over the whole Earth. These results were presented by CW in July 2001 on behalf of our team at the *Instruments, Methods, ad Missions for Astrobiology IV* session of the *SPIE* Meeting in San Diego. The presentation made international headlines. Doubts about contamination were naturally raised by sceptics imbued with a geocentric worldview, but our initial results have since received extensive confirmation in later work including recent studies by a group of scientists in Japan.

While the work carried out in Cardiff failed to grow microbes from the stratospheric collections, one of us (MW) succeeded in culturing two separate bacterial species that were related to known terrestrial species. Although this work was published in a reputable microbiology journal that was subject to rigorous peer review, it should be placed on record that hostile comments sprang from many sources. One comment stating that the cultured microbes, despite all the controls that were used, had to be contaminants for the reason that they were similar to terrestrial species shows that our critics have not read the book by Hoyle and CW entitled *Evolution from Space*. If microbes on Earth are derived from comets, and continue to be replenished on the timescale of tens to millions of year, then it is to be expected that new organisms will be similar to ones that are resident on Earth. Minor evolutionary genetic drifts are all that are expected, and these are indeed what are found.

Perhaps the most noteworthy intervention as stated was a letter written by Nobel Laureate Sir Martin Evans to a leading Indian Scientist, Pushpa Bhargarva denouncing MW's work as being "not worthy of a first year undergraduate project report"! So deeply ingrained was the prejudice against this work at this time, and sadly this has not changed much in 2022.

Another project that we were drawn into somewhat unexpectedly was an association with the Indian scientist Godfrey Louis to investigate a remarkable phenomenon — the Red Rain of Kerala. In the summer of 2001 red rain fell in large quantity over much of the state of Kerala in south India, and Godfrey Louis, a physicist at Mahatma Gandhi University, conducted a series of laboratory studies on samples of the rain that he had collected. His initial results showed that the red colour of the rain was due to the presence of pigmented biological cells. They were generally similar in appearance to algal cells but they were unlike any algal cell known thus far. Godfrey Louis claimed that the cells could be cultured when they were placed in a hydrocarbon-rich medium under high pressure at a temperature of 450°C — higher than the survival temperature of any known bacterial or algal cell. The fall of the Kerala rain had been preceded by a loud sonic boom, and it was conjectured that a small fragment of a comet exploded in the stratosphere, unleashing vast quantities of red cells that became the nuclei of raindrops. This intriguing story had an obvious possible link to cometary panspermia theories, so it was natural for Godfrey Louis to seek a collaboration with us. A sample of the rain was sent to Cardiff and we worked on this for nearly a whole decade. It may sound surprising, but it remains the case that to this day we have not been able to identify these cells. The biological nature of the red cell material was confirmed by Milton Wainwright and independently by a PhD student Gangappa Rajkumar, and also its replication under high pressure to a temperature of 121°C. There have been claims that Red Rain cells lack DNA, although MW and co-workers showed that they, in fact, stain positive for DNA with the DAPPI stain. Figure 7.4 shows Red Rain cells; note the presence of a flagellum on the top-most cell and a distinct red, caused by the vacuum of the scanning electron microscope, emphasising that the cells have thick walls.

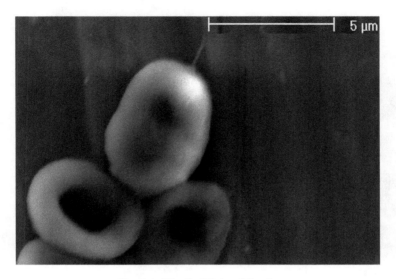

Fig. 7.4. The Red Rain of Kerala.

In November 2012 red rain episodes similar to what happened in Kerala in 2001 were reported over large areas of central Sri Lanka. Investigations of the Sri Lankan red rain have shown that the red cells present in this rain are very similar to the unidentified cells of the Kerala red rain that we already discussed. A significant difference, however, was the nearly simultaneous reports of fireball sightings and a fall of carbonaceous meteorites, strongly suggesting a causal link. Unfortunately, red rain remains an enigma, one which will not be solved until the scientific community engages with the phenomenon and significant research is conducted in the origin and physiology of these cells.

8

The Microbiology of Panspermia

The theory of panspermia is becoming increasingly more sophisticated as new variants of the theory are being developed. It is necessary therefore to define the various "sub-species" of the theory that are currently on record. Here, we shall restrict the term Panspermia to incorporate the view that life on Earth necessarily had a cosmic origin. A second variant, so called Neopanspermia will be used to refer to the view that life, not only arrived to Earth from space, but continues to do so, while the term Pathospermia will be used to cover the theory that diseases occurring on Earth, such as influenza and SARS and COVID-19 originate from space. Other variants of panspermia include the view that microbes were brought to Earth in meteorites, so called Lithopanspermia, and even that they may one day be ejected from terrestrial building material flung out into space from Earth impact events, so called Archaepanpsermia.

From its very beginning, the primitive Earth, we would suggest, would be exposed to life in the form of proto-life, small organisms, and naked DNA — all securely protected from the harsh conditions of space by a covering of cosmic dust or by being conveyed within a bolide that perhaps would include frozen ices. Such life forms would tend to be consumed by the hot "fireball" conditions that prevailed on the early Earth; however, at the point of cooling such life forms could gain a footing, survive on the early planet and proliferate. It is likely that these life forms would arrive as a mixed population of space-derived heterotrophs and autotrophs, which would immediately compete with each other. Due to the law of doubling and because of the lack of competing organisms on the proto-Earth, the first incoming organism(s) would rapidly cover and colonise the entire planet. Both autotrophs and oligotrophs (i.e. those heterotrophs which can live on trace of nutrients) would initially be at an advantage, but the continued input of organic matter from space would quickly support heterotrophs. Any organic material protecting the incoming life forms would also likely provide nutrients. Incoming viruses would exist independent of such nutrients, but would need the development of a host; incoming viruses would therefore be associated with a host or need to acquire one. The establishment and development of these individual life forms would depend on the prevailing conditions and their establishment would vary as a result. Waves of different populations would cover the newly cooled plant and would vary as the conditions varied with the survival of the fittest prevailing. Life might have originated by the chemical theory had this input of space-life not occurred, but any proto-life forms of terrestrial origin would soon be preyed upon by the alien forms. The export, from Earth to the cosmos, of newly established forms could possibly follow the myriad of impact events on the early planet — a kind of negative panspermia.

The above suggested sequence of events is, however, unsustainable without the presence of an atmosphere. This is not just to sustain any organism requiring gases such as oxygen or carbon dioxide, but because a newly formed atmosphere would slow down incoming life form-delivery systems and prevent them being annihilated as they smash into the Earth's surface. As a result, panspermia-derived life would have difficulty in establishing itself before the formation of an early Earth atmosphere.

It is often pointed out that there was insufficient time for life to have evolved on Earth, a problem which would not limit the arrival, and establishment here of proto-life, or organisms from space. The probability of life arriving from space and establishing on Earth will be considered further below, but it is important to note that vast amounts of this material continually arrive on Earth, as they would have done in the past, and the establishment of life would only require one organism to survive and continue to replicate in a new environment.

The Space-Life Carriers

The extreme conditions existing in space means that it is unlikely that DNA, proto-life and developed life would arrive on Earth in the naked form. Instead they would need to be protected by some form of covering to protect them from the lethal effects of UV (notably UVC) ionizing radiation and extremes of temperature, etc. The best form of such protection would be provided by conveyance in a meteorite, asteroid or comet, but could also be provided, more simply, by a covering of cosmic dust. Hoyle and Wickramasinghe suggest a layer of carbon i.e. soot could protect life forms, as could cosmic dust. All of such carriers would be obliterated in the absence of an atmosphere, which explains why contemporary moon dust does not contain panspermic organisms. Even in the presence of an atmosphere, most incoming organisms would burn up, but sufficient number would survive to inoculate the planet. Comets are likely to provide an excellent vehicle for the transport of life forms from space to Earth.

The Cosmically Inoculated Consommé

Ever since Darwin came up with the idea of a warm little pond scientists have tended to believe in the idea of a primordial soup. Readers will be aware by now that we enjoy delving deep into the historical literature to find earlier discoverers of ideas often erroneously attributed to more famous people; yet we can find no evidence that anyone before Darwin came up with the "warm, little pond" metaphor. Does this metaphor conceivably have a relevance to the modern theory of panspermia? It probably applies at some distant epoch — Hadean epoch some 3.5–4.3 billion years ago seems the best bet. Then the Earth's surface had cooled but there was still a relentless protracted episode of comet impacts which brought water to form the Earth's oceans. Evaporation of water from the oceans led to the formation of an atmosphere, so conditions became optimal for life to take root. However, the theory of panspermia states that, rather than it developing here on Earth by unproven processes, it arrived from space in the comets that were then bombarding Earth. Comets, as we discussed earlier, are the natural repositories and incubators of panspermic microbiology. This input from comets thus made up the inoculum that would grow and proliferated in the consommé of nutrient already present on the Earth's surface. This inoculum could have been made up of single organisms or proto-organism or, more likely, a mixture of many different types. It might also have contained representatives of one or a *pot pouri* of the major metabolic groups, such as aerobes, anaerobes, heterotrophs, chemotrotrophs, and photo-trophs (i.e. photosynthetic organisms). In short it would likely be a mixed culture of microbes.

The consommé would contain all the nutrients needed by the incoming microbes, including carbon sources, nitrogen as well as a variety of mineral nutrients. The term consommé is preferred to soup because the Earth-bound medium is likely to have been dilute and not a thick soup. This would have given oligotrophs, that is microbes which can grow by scavenging trace amounts of nutrients, an initial advantage. There would be continual battle for dominance between arriving organisms (which would already have been exposed to competition on the parent comet). And of course there would have been successive waves of dominant organisms with individuals beginning to exploit niches; some would be aquatic, others terrestrial, and their dominance would rely on standard factors such as nutrient availability, temperature and pressure.

As soon as oxygen becomes available (via evaporation of water from the oceans, and dissociation of water molecules), aerobic heterotrophy would begin to dominate, since this is the most efficient way in which a microbe can gain energy. It goes without saying that the organisms would compete with each other and that evolution would be kick stated. It would only require a single organism or proto-organism to arrive for life to become established. Within a short time, this single organism, let us call it the "pansperm", would, due to the law of doubling cover the planet. Once a microflora becomes established then it would provide a food source for other organisms delivered in the vast quantities of water supplied by comets. The term water is a misnomer, since it would not be pure water but a mixture of carbon, nitrogen, and other potential microbial nutrients. So, within a very short time a living biomass would cover the Earth and be under a process of continual change and evolution. New organisms would then come in over the eons as they themselves evolve in comets. This would be reduced as comets cease to impact the Earth, but microbes, in the form of biological entities, or even bacteria and fungi (moulds) will be deposited following cometary ablation. The inoculum would also contain DNA and other biological material, which will add new information, and contribute to evolution on Earth, even unto today. Here then lies the potential for life to evolve, following the arrival of a variety of incoming components that become assembled rather like a child's jig-saw puzzle. Thus, for example a delivered protoplast cell might pick up a delivered mitochondrion or nucleus by the process first suggested by Schimper and Meraschowski and more recently developed by Lynn Margulis. Impact events on Earth might then also subsequently spread these organisms back into the cosmos, by a process of reverse panspermia. Later impact events over aeons would also destroy some organisms, allowing others to develop, such that even if a comet or asteroid impact were to destroy all life on Earth, the planet would soon become re-inoculated, although the long-term outcome would likely be very different. This process will have occurred with differing outcomes on any planet on which conditions allow for microbes to develop. The prevailing conditions may allow for the selection of only one metabolic group, or species of delivered organisms. For example, an aerobe might alone be taken up by a planet lacking oxygen. Should photosynthetic organisms get a foothold on such a planet then the resultant release of oxygen and decrease in carbon dioxide would allow new inoculants to gain a footing and obviously change the planet forever. This planet-modifying ability of microbes is of course critical, once developed microbes could be transferred from one plant to another, as in the possibility that life on Earth was transferred from Mars.

Microbes and Panspermia

Life might have originally arrived on Earth from space as a single organism (or proto-organism) or, perhaps more probably, as a mixed population made up of heterotrophs (aerobes, anaerobes and aerobic oligotrophs), phototrophs and chemoautotrophs). The survival of organisms possessing these metabolic strategies would depend on the conditions they meet, such that their survival and establishment would be selected by the prevailing conditions on Earth.

However, if the panspermic organisms arrived in a meteorite they could bring their own microenvironment with them in which a mixture of, say, cyanobacteria and aerobic heterotrophs might arrive and survive together, the former (following exposure to sunlight) providing oxygen to support the latter. In this case colonisation of the Earth need not, initially, be widespread but may have occurred locally in the face of general, unfavourable physical conditions. A lone aerobic heterotroph that might have arrived on Earth devoid of oxygen would obviously not become established; however, if it arrived along with an oxygen-generating phototroph it might well have survived on an otherwise oxygen-free, but carbon dioxide-rich Earth. Perhaps diatoms might act as effective "life capsule" for the transfer of life to Earth as their silica shell would likely protect the internal phototrophic protoplast from environmental extremes, notably UV radiation. It is known that bacteria can co-exist within diatoms cases, so it is possible that a heterotroph could be supplied to Earth in a rain of diatoms and survive even in an oxygen-free environment, ready to be released from its diatom shell when the Earth's oxygen supply improves.

Survivability of Microbes in Extreme Cosmic Conditions

When Lord Kelvin suggested that life might have originated from space, critics quickly commented that there was no way that living things could possibly survive the transit from the outer regions of the cosmos to Earth. At that time, microbiology, the study of algae, bacteria, microscopic fungi and protozoa, was in its infancy, and little was known about their survivability While the critics might have been right about the difficulty of transferring higher organisms, we now know that microbes are amazingly hardy, and recent studies have shown that they are likely to be able to survive transfer through space and the resultant impact of landing. It should also be remembered that incoming microbes, in many cases, be protected by being contained within protective meteorites and dirty cometary ice.

Many terrestrial microorganisms can live through periods of water and nutrient deprivation over long periods of time. Some bacteria can produce thick-walled protective endospores, or become dormant by closing down almost all of their metabolic machinery. In relation to survival over time, dormant bacterial spores from insects embedded in amber for 40 million years can be resuscitated, while a halotolerant *Bacillus* (a bacterium which produces endospores) has been apparently isolated from a 250 million years old salt crystal. Microbes occurring in ice cores obtained from the bottom of Lake Vostok in Antarctica may serve as a model simulating conditions inside the permafrost subsurface place of Mars or Jupiter's moon Europa. In contrast, hyperthermophilic (extreme heat-loving) microbes found on Earth may have analogues on warm planets such as Venus or the volcanically active moon, Io. Such facts suggest the distinct possibility of viable cometary transfer, with microbes surviving inside the melted centres of icy comets. Microbes can even survive a high vacuum and when exposed to severe galactic radiation. Microorganisms also possess remarkable DNA-repair mechanisms. The extremophile, *Deinococcus* can also survive sterilising solar photons in space. UV radiation, notably UVC is particularly damaging to microbes but bacteria such as *Bacillus subtilis*, survive exposure to UV while exposed on Earth orbit, merely by being shielded by only a single layer of dead cells. Even if only one spore out of 10,000 can avoid lethal cosmic rays for hundreds of thousands of years, this survival rate would be significant in relation to panspermia. High temperatures in space might also be expected to kill microbes, but it has been confirmed that dried microbes and spores can survive high pressure and heating to 350°C for 30 seconds. An indication of the survivability of bacteria in the space environment is indicated by the fact that *Streptococcus mitis* was isolated from a camera on board Surveyor 3 after spending nearly three years on the Moon.

Necropanspermia, Necrosymbiosis and the Origin of Life

To date, ideas about panspermia have concentrated on the arrival to pre-biotic Earth of proto-life forms, living bacteria and possibly other microorganisms which then grow and evolve into present day life. Another speculative possibility is that bacteria do not survive the journey through space, but dead, freeze-dried biomass arrives instead; what might be termed "necropanspermia". This dead biomass would add the fully-formed components of life and bacterial degradation products to the prebiotic Earth.

Proto-life that has developed here could then take up these essential components like DNA, cell membranes and mitochondria and develop them into fully formed life, a kind of necrosymbiosis would thereby occur. In this way, many of the complicated stages in the development of life on Earth would be circumvented. Possibly, this necrosymbiosis would occur in drying pools, as the preformed components of life were brought into intimate contact. It would certainly be interesting to determine if cellular components can be taken up and be activated by simple life forms, although it could be argued that these are already far too complicated for this Frankenstein-like origin of cellular complexity to occur.

9

Evidence for Fossilized Life in Meteorites

A number of scientists now believe that microorganisms can be transferred between various planets including the Earth and their moons in our solar system, and by extension in the many other exoplanetary systems that are now known to exist. It is possible that such "negative, or reverse panspermia" could have seeded life from this planet to Mars, or *vice versa,* or even transferred life across many hundreds of light years between exoplanetary systems. While panspermia has long been hypothesized and recently developed with great force by the present writers and Fred Hoyle amongst others, there has simultaneously been a recent surge of enthusiasm for the theory, following claims of potential extraterrestrial fossils in meteorites. Of particular note is the (oft disputed) evidence for fossilized microbial life in the ALH84001 meteorite.

Some twenty-four thousand meteorites are reckoned to have reached the Earth, around one hundred and thirty having originated from the Moon and over thirty from Mars. A number of these meteorites had never achieved temperatures above 100°C, and were therefore never exposed to temperatures high enough to sterilize all parts of their interiors. An incoming body large enough to punch its way through the atmosphere (where it exists) of one of the terrestrial planets may strike the ground at a few several kilometres a minute, and produce a crater thus ejecting stones and soil at high speeds.

Calculations suggest that the portion of Earth material ending up on Mars following a 10-million-year period of impacts would be around 0.16%. Even a gram of basalt may contain 10^7–10^8 microbes within fissures, making possible the viable transfer of microbes between Earth and nearby planets. Such possibilities rely on whether lifeforms are able to survive the "stun-effect" of being ejected into space. Scientists have confirmed, using impact studies, that large numbers of microbes are likely to survive such discharge and planetary impact on re-entry. Ejected material may also ride on the vapour plumes, making it simpler to accomplish escape velocities. Finally, it has also been suggested that the debris produced during supernova formation can carry living spores as some of this debris falls to Earth.

Darwin and Meteorites

There is an amazing claim that Darwin was shown what appeared to be fossilized life inside a meteorite and reacted with joy and amazement. The meteorite was apparently shown to him by a German professor, Otto Hahn (Fig. 9.1). Hahn's work seemed to show that life originated on Earth from space, and the story goes that Darwin became a believer in panspermia. How did this all come about? Well, it first appeared in the August 1881 issue of the New York magazine, *Science* and was for a long time overlooked. The article was entitled, *Mr. Darwin on Dr. Hahn's Discovery of Fossil Organisms in Meteorites. Science* was founded the preceding year by journalist John Michels, and was financed by Thomas Edison (who, as we have already seen believed in panspermia and would doubtless have been happy promoting the story). The current top-rated journal of the same name is a later version. The article contains some amazing quotations from a lost Darwin letter or letters and, even more surprisingly, spoken words attributed to Darwin upon viewing meteorite specimens. The passage reads:

Fig. 9.1. Otto Hahn

Not content with the mere presentation of his work, Dr. Hahn visited the veteran zoologist [Darwin] and brought his preparations to him for inspection. No sooner had Mr. Darwin peered through the microscope on one of the finest specimens when he started up from his seat and exclaimed: "Almighty God! What a wonderful discovery! Wonderful!

And after a pause of silent reflection he added: "Now reaches life down!"

Hahn (1828–1904), was a lawyer, author, and amateur petrologist in Reutlingen, Baden-Württemberg, in the newly-formed Germany. In his most celebrated and controversial work: *Die Meteorite (chondrite) und Ihre Organismen* he claimed to have found microscopic, fossilized sponges, corals and crinoids inside meteorites. The book was well illustrated with thirty-two micrographs. On 16 December 1880, Hahn forwarded a letter to Darwin together with his new book.

At first it seemed like Otto Hans' work would get a sympathetic hearing from the scientific community. However, this acceptance failed to last, and after 1882, Hahn published no more about fossils or meteorites, and in an 1889 autobiographical sketch he barely mentioned his meteorite and fossil work, seeming as if he no longer wished to be associated with it. Hahn died in 1904 while on a trip back to Germany from Canada, where he had emigrated.

The Hahn–Darwin story has been extensively quoted. Unfortunately for advocates of panspermia, the Darwin scholar Jan de Whye has concluded that this story is just a myth and that we can be reasonably sure that Darwin did not believe in Hahn's claim that life on Earth came from outer space. Van Whye further concludes that:

> Given the consistency of Darwin's statements on the origin of life, the claim that he believed in an extra-terrestrial origin seem extremely implausible.

Henri Becquerel and Meteorite-Microbes

According to the physicist, Henri Becquerel, Louis Pasteur was one of the first to conduct investigations into the possible occurrence of bacterial spores in stony meteorites. However, a later study by Gallipe and Souffland in 1921 apparently proved negative. Then, some ten years afterwards C.B. Lipman claimed to isolate a number of different bacteria inside surface-sterilized meteorites. These findings not surprisingly led to intense controversy, although Lipman's work was repeated by Roy in 1935 when he isolated bacteria from meteorite samples in three out of twelve experiments; he identified the isolates as common species, including, *Bacillus subtilis* and *Staphylococcus albus*.

More recent attempts to culture living micro-organisms from meteorites include those of Briggs and Kitto in 1962 who were unable to detect viable organisms. However, in the same year, the presence of another unidentified aerobic bacterium in a meteorite was claimed by

Frederick D. Sisler. Incidentally, Sisler invented a biochemical fuel cell principle, a process that uses bacterial action to extract electric power from organic waste. In the same year, the Russian microbiologist, Rubchikova again grew unidentified micro-organisms from the interior of yet another carbonaceous meteorite.

In later years scarcely any attempts have been made to isolate microbes from meteorites. The general consensus is that such isolations prove nothing, since critics can simply say that any such isolated organisms are terrestrial contaminants, no matter how strenuous are the efforts to use sterile isolation techniques. It has been suggested that the only way that the presence of microbes in meteorites would be deemed credible is if a totally new and exotic organism unknown to Earth were to be isolated. Even so it would be argued by sceptics that such an organism already existed but had yet to be isolated from terrestrial samples. The statement made by S. Abyzov and A. Imshenetsky, that:

> .... meteorites extracted from soil not at once after their fall are hardly suitable for microbiological assay....

epitomises this attitude.

As a consequence, it is clear that it would obviously be desirable to intercept a meteorite, or meteorite dust in space above the terrestrial biosphere, before it could get contaminated by terrestrial microbes. If such material could be shown to contain microbes, then this might just serve to convince sceptics of their cosmic origin.

The most famous claim for fossils existing in meteorites relates to studies on the Orgueil carbonaceous chondrite, discovered in 1864. On May 14, a meteor shower fell near the town of Peillerot in southern France. The meteorites, which were composed of carbonaceous chondrite, were given the name 'Orgueil', and distributed throughout Europe for study. It is said that "Louis Pasteur investigated it, but was not impressed, a story which may be apocryphal; since then there have been over four hundred publications on the stone alone. It is remarkable how many papers were published during the 1960s, in the prestigious journal, *Nature*, claiming that carbonaceous meteorites contain fossilized life. In 1966, the Nobel Laureate, Harold C. Urey extensively reviewed the evidence for biological materials in meteorites, observing that the organic substances present are unlike recent rock contaminants.

Critics of the idea that the Orgueil meteorites contained fossils (so-called "organized elements") suggested that what supporters of the idea were seeing were pollen grains, a claim dismissed by the pollen expert, Martine Rossignol-Strick, and the Archaean microfossil pioneer, Elso Barghoorn.

The Orgueil meteorites remained all but forgotten until the early 1960s, when Bart Nagy examined samples of them and found curious microscopic intrusions that resembled fossils. His work published in *Nature*, began a debate that continues to this day concerning the presence of fossils in meteorites. A team of Chicago researchers then found plant fragments (entire seeds) and coal embedded deep inside one of the meteorites that had been sealed inside a glass jar and stored in the museum at Montauban. Not surprisingly, it was immediately suspected the plant and coal fragments had somehow contaminated the meteorite and were indigenous to it. But x-ray analysis surprisingly, ruled this suspicion. The plant fragments were definitely embedded in the meteorite itself. Additionally, the entire meteorite was encased in an impenetrable glass-like fusion layer created by heat as it passed through the atmosphere, pointing to intriguing possibility that the plant seeds were extraterrestrial in origin. Critics then suggested someone had wet the rock and inserted the plant fragments into it, where they remained as the meteorite. But what about the glassy fusion layer? Further tests on the rock revealed that this fusion layer was in fact dried glue, although pieces of the original fusion layer remained within the meteorite.

It seemed obvious that sometime around 1864, before the meteorite had been sealed inside a glass jar, someone had embedded plant and coal fragments inside of it, and then coated it with glue to try and recreate the fusion layer. Why had someone gone to the trouble of perpetrating this hoax? One suggestion is that it was all a joke, aimed at discrediting the idea of spontaneous generation, an idea that was in vogue at the time. The joke backfired however, since no one noticed until nearly hundred years later in 1962!

Even though one of the Orgueil meteorites had obviously been tampered with, this did not have any bearing on whether or not the Orgueil meteorites contains microfossils. That debate rages on. However, if there is one thing that destroys an idea in science, it is a hoax. No matter that the hoax was not relevant to the main claims made about fossils, it still provides a reason for critics to summarily dismiss anything relating to the Orgueil meteorite and interstellar life.

The whole argument was revived in the early 1960's by Harold Urey, together with G. Claus, B. Nagy and D.L. Europe. They examined the Orgueil carbonaceous meteorite, that fell in France in 1864, microscopically as well as spectroscopically. They claimed to find evidence of organic structures that were similar to fossilised micro-organisms, algae in particular. The evidence included electron micrograph pictures, which even showed substructure within these so-called "cells". Some of these structures resembled cell walls, cell nuclei, flagella-like structures, as well as constrictions in some elongated objects that suggested a process of cell division. These investigators, like their colleagues before them, became immediately vulnerable to attack by orthodox scientists. With a powerful attack being launched by the most influential meteorite experts of the day, the meteorite fossil claims of the 1960's became quickly silenced.

From the early 1950's a fierce debate took place in the column of scientific journals as to whether or not there was evidence of fossilised microorganisms in carbonaceous meteorites. Although evidence for the existence of such "microfossils" remained strong, the proponents of "no evidence" seemed to win the debate for reasons that were not entirely based on science. However, as always, the correct ideas are never silenced for ever. In the late 1970's the palaeontologist and microscopist, Hans Dieter Pflug from the Geological Institute of Justus Liebig University in Giessen, Germany entered the fray with startling new evidence that could not be easily refuted. In 1979 Pflug published evidence for microbial fossils in the sedimentary rocks of South-West Greenland (the Isua Series) that already caused a stir. These rocks being dated at 3800 million years put the first appearance of life back by some 500 million years from previous estimates, thereby reducing the time available for the development of any primordial soup. This work has been later supported by studies of several other researchers who have pinpointed the oldest evidence of life on Earth to be at a time close to 4.2 billion years ago, at a time of intense bombardment by comets.

Hans D. Pflug and Meteorite Microfossils

In 1980 Hans Pflug (Fig. 9.2) contacted one of us (CW) offering information that was even more interesting than evidence of the oldest terrestrial microfossils. He claimed to have discovered compelling new evidence of bacterial microfossils in carbonaceous meteorites.

Fig. 9.2. Professor D.H. Pflug

As the name implies, the carbonaceous meteorites, contain carbon in concentrations upwards of two percent by mass. In a fraction of such meteorites the carbon is known to be present in the form of large organic molecules. It is generally believed that at least one class of carbonaceous meteorite is of cometary origin. If one thinks of a comet containing an abundance of frozen microorganisms, repeated perihelion passages close to the sun could lead to the selective boiling off of volatiles, admitting the possibility of sedimentary accumulations of bacteria within a fast-shrinking cometary body. We can thus regard carbonaceous chondrites (a type of meteorite) as being relic comets after their volatiles had been stripped.

In 1980 Pflug reopened the whole question of microbial fossils in carbonaceous meteorites with great force. Using techniques that were far better than those used by earlier workers, he found a profusion of cell-like structures comprised of organic matter in thin sections prepared from a sample of the Murchison meteorite which fell in Australia, about a hundred miles north of Melbourne on 28 September 1969. He showed these images to Fred Hoyle and CW and they were immediately convinced of their biological provenance. Pflug himself was a little nervous to publish these results, fearing for his career and anticipating the kind of reaction that had been seen in the 1960's. His work latest results were however presented at an out-of-town meeting of the Royal Astronomical Society, held in 1980 in Cardiff (Fig. 9.3).

Pflug's method was to dissolve-out the bulk of the minerals present in a thin section of the meteorite using hydrofluoric acid, doing so in a way that permits the insoluble carbonaceous residue to settle with its original structures intact. It was then possible to examine the residue in an electron microscope without disturbing the system from outside. The patterns that emerged were stunningly similar to certain types of terrestrial microorganisms (Fig. 9.4). Scores of different morphologies turned up within the residues, many resembling known microbial species. It would seem that contamination was excluded by virtue of the techniques used, so the sceptic has to turn to other

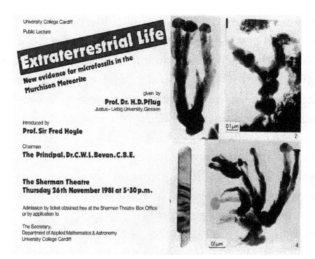

Fig. 9.3. Pflug's lecture poster and an example of a microfossil of *pedomicrobium* in the Murchison meteorite.

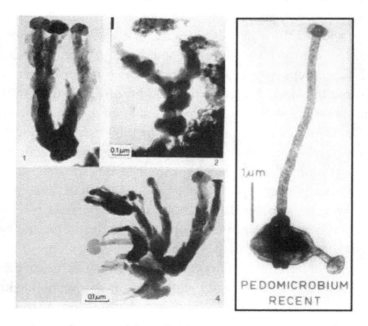

Fig. 9.4. Pflug meteorite-morphologies resembling modern microbes.

explanations as disproof. No convincing non-biological alternative to explain all the features was readily to be found.

Perhaps the most important and exciting recent research on the presence of fossilized life in meteorites has been done by the former NASA scientist, Richard B. Hoover. Hoover has been providing convincing evidence of the presence of fossils in meteorites for a number of years. Yet again, the scientific establishment has ignored his studies and he has been disgracefully disowned by NASA. Richard's work emphasises the presence of Cyanobacteria (bacteria which used to be called blue green algae) in fossils and as we shall see, we have independently verified some of his findings regarding these fossilized organisms.

10

Our Evidence for Meteorite Microfossils

Meteorites are fragments of rock originating from either asteroids or comets that populate interplanetary space and which survive impact on the Earth's surface; a fragment of such material is referred to as a meteoroid before its impact. The study of meteorites can reveal how the solar system first began, what life (if any) is like on other planets and how such planets were formed. Rocks originating from elsewhere in the solar system therefore carry with them crucial information about the planets or planetary bodies from which they originated. Meteorites could also potentially bring evidence for the past existence on Earth of other life forms that are now extinct.

Without doubt the most famous recent claim that meteorites contain fossilized microbes involves the meteorite-ALH84001 which was discovered 1984 in the Allen Hills region of Antarctica. Electron microscope studies showed that this meteorite contains very tiny grains of a magnetic–mineral made of oxygen and iron surrounding tubes of width 20 nm and a length of 100 nm, which looked like nanobacteria (i.e. ultra-small bacteria). A group of scientists at NASA led by David McKay, claimed to have found evidence of bacteria having lived on the meteorite when it existed as a rock on the surface of Mars. This meteorite has become the most studied rock in recent history, yet still we cannot confirm that these extremely small, fossilised objects are in fact bacteria.

As we have already noted, convincing evidence of microfossils, similar to Cyanobacteria, was discovered by Richard Hoover who claims that these fossilized bacteria were once living organisms in the parent bodies of the meteorites — asteroids, planetoids, moons, comets — and are not Earth contaminants. Both authors have similarly, independently provided evidence for the presence of fossilized diatoms and other microbes in meteorites. To date, none of these claims have been fully accepted by the scientific community, largely on an excuse that meteorites contain mineral features that closely mimic bacteria and other microorganism thus making it difficult to distinguish between potential fossilized bacteria and non-biological artefacts. And yet again, the curse of contamination comes into play.

Having read recent studies on the claimed presence of microbes in meteorites, one of us (MW) obviously could not avoid the temptation to look for life within these bolides. Again with the help of Alex and Chris, MW began to work with a sample of the Northwest Africa 4925 meteorite. We chose this because it is commercially available and confirmed to be *bona fide* meteorite. The single sample supplied from a dealer was cut from an originally larger meteorite sample, which was catalogued as Northwest Africa 4925 NWA 4925; members of the International Meteorite Collectors Association confirmed the authenticity of the sample. The meteorite was recovered in 2007 from Erfoud, a town in the Sahara Desert, in the Meknès-Tafilalet of the Maghreb region in eastern Morocco. (Fig. 10.1). The fragment which was used was covered partially with a fusion crust, and showed a porphyritic texture with large chemically zoned olivine megacrysts set into a fine-grained groundmass composed of pyroxene and maskelynite; minor phases include chromite, sulphides, phosphates, and small Fe-rich olivines. Olivine megacrysts often contain melt inclusions and small chromites. Its mineral composition (EMPA) is as follows: Olivine, Fa27.6–46.8; pyroxene, Fs20.0–37.7Wo3–14.8; maskelynite, An67–69. It is classified as an achondrite (Martian, olivine-phyric shergottite); severely shocked with some melt pockets and moderately weathered.

As has already been mentioned, when meteorites fall on Earth they rapidly become contaminated, at least on the surface, by terrestrial

Fig. 10.1. The commercially sourced, North West African Meteorite.

microorganisms. So, in order to perform any microbial studies on them, we must first sterilize the surface, so as to kill and lyse (break apart) any microbial cells. To achieve such sterilisation, meteorite samples were immersed in seventy percent ethanol for an hour, and then washed twice with twice sterilised, deionized water, before being transferred to a sterile Petri dish prior to being examined under a SEM. (Further details of the methods employed can be found in the Appendix).

Again MW and his team were surprised and excited by what they found, but it would be as nothing compared to what we would see in later studies. Figures 10.2 and 10.3 show the inside surface of the sample after being carefully scanned and imaged. It shows a presumptive fossilized bacterial biofilm made up of typical bacterial forms, which would have originally been located within the original meteorite from which the sample was cut. Had this so-called biofilm been detected on the surface of terrestrial soil or rock samples (either living or fossilised) most microbiologists would have no problem in describing these small forms as being bacteria. Biofilms are organised aggregations of microorganisms living within self-produced extracellular polymer attached to an inorganic surface; by living in such communities, microbes are protected against stresses such as antibiotics and the destructive effects of UV.

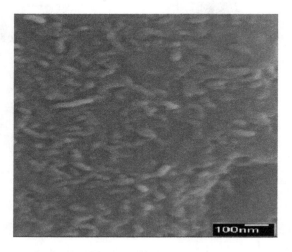

Fig. 10.2. A presumptive fossilized bacterial biofilm inside the Northwest African meteorite.

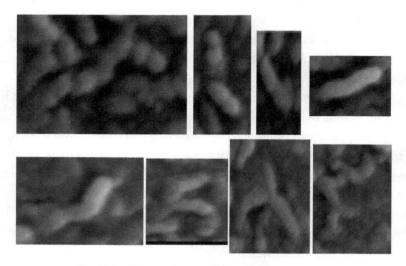

Fig. 10.3. Close up image of bacteria-like structures.

Figure 10.3. shows a magnified image of the meteorite biofilm that appears to be individual bacterial types with obvious cocci, rods, individual rods and spiral shapes bacteria-like chains. The chemical composition identified by EDAX of (a), the presumptive "biofilm" region, and (b) an area away from the "biofilm" (non-biofilm region) are essentially similar, being comprised of iron, magnesium, calcium, silicon and oxygen (Fig. 10.4).

The sample has clearly been cut from a larger piece so that the outside surfaces would originally have been well inside the meteorite. Critics could further argue that the observed bacteria-like structures are simply mineral artefacts simulating the morphology of bacteria, which of course would prove to be a remarkable coincidence. The bacterial forms seen in the biofilm here, unlike those seen in the Allen Hills meteorite (which are nano-sized), are around 0.2 μm check (i.e. similar in size to terrestrial bacteria found in natural, nutrient-limited environments found on Earth; and again, in comparison to the Allen Hills form, the claimed bacteria described in our work are large enough to contain a complete bacterial genome.

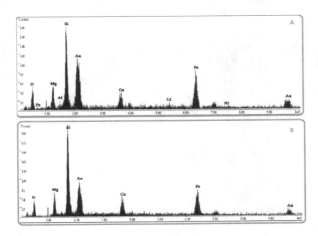

Fig. 10.4. Two almost identical EDAX readings for different regions within the same sample. Northwest Africa 4925 Martian meteorite, (A) is for the "biofilm" region, and (B) is for another area within the broken sample which is away from the "biofilm" region.

It might be suggested that the image is of a recent bacterial biofilm which presumably formed after the meteorite landed on Earth. However, the fact that the bacteria having been exposed to 80% ethanol, show no sign of lysis suggests otherwise; a modern terrestrial biofilm might have undergone mineralization during the period when the meteorite resided on Earth, but this seems highly unlikely. Similarly, it is highly unlikely that a modern terrestrial biofilm could have formed from an air-derived bacterial inoculum while, the meteorite was kept in storage. Yet again, we would doubtless be accused of suffering pareidolia, the curse of seeing what you want to see.

Finally, Fig. 10.5 shows fossilized filaments in the Mars meteorite which are almost identical to those found in other meteorites by

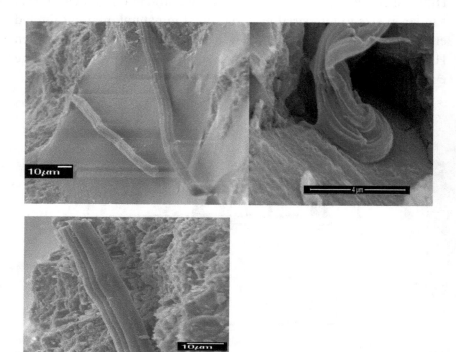

Fig. 10.5. Fossilized filaments in the Mars meteorite identical to those found by Richard Hoover.

Richard Hoover and identified by him as Cyanobacteria. Either different meteorites contain fossilized Cyanobacteria, or they contain similar mineral artefacts. Finally, a French group, led by P.H. Gilet recently reported finding fossilized bacteria in the Tatahouine meteorite which, as in the Allen Hills findings, were extremely small; much smaller than the bacteria we describe finding.

We shall next discuss finding remarkable biological objects within a meteorite, inclusions that go well beyond what we see above. Needless to say, our critics will continue to find a way of summarily dismissing our work, without really providing any detailed scientific explanation or criticism of why we are wrong; a pattern of response that we will become well used to!

Amazing Fossils in the Polonnaruwa Meteorite

On 29, December, 2012 a large meteor that appeared in north-central Sri Lanka was seen to fragment in the sky into many pieces and a spray of meteoroids fell on a rice paddy field located a few miles away from the historic ancient city of Polonnaruwa. Figure 10.6 shows the location of the fall and a small piece of the meteorite.

The Polonnaruwa meteorite shown in Fig. 10.6 (right) exhibited many characteristics of a carbonaceous chondrite but with a remarkably porous structure. This meteorite on account of its timing was provisionally identified as arising from a cometary fragment in the Taurid complex associated with comet Encke. A few percent carbon as revealed by EDX analysis made it consistent with the properties of carbonaceous meteorites but its exceptionally low density (0.6–1.2 g/cm^3) and high porosity made it significantly different from hitherto classified carbonaceous meteorites, and hence its acceptance as a meteorite has come to be challenged. Not only did critics claim that such low density asteroidal/cometary bodies do not exist, but it was subsequently argued that structures of this type could not survive entry through the atmosphere.

Left Fig 1a Map of Sri Lanka
showing location of meteorite

Right Fig 1b Piece of meteorite

Fig. 10.6. Location of fall and sample of the Polonnaruwa meteorite.

Over the past few years these reservations have been shown to be unjustified and the meteoritic/cometary origin of the Polonnaruwa stones appears to be vindicated. Analysis of short-lived uranium isotopes in carbonaceous chondrites has yielded excesses of 234-uranium over 238-uranium, and 238-uranium over 230-thorium from which it could be concluded that a fluid condition must have persisted in the parent bodies of these meteorites as recently as during the past few 100,000 years.

There is much controversy over whether the Polonnaruwa stone is a true meteorite, a fact which is not helped by the Wikipedia article about it being totally inadequate and omitting any recent studies on its composition and providence, leading the reader to conclude that it originated on Earth. It is claimed for example, that the Polonnaruwa meteorite is really a piece of what is called fulgurite. Fulgurite is commonly referred to as "fossilized lightning", and is made up of tubes, clumps, or masses of sintered, and glassified soil, sand, rock, organic debris and other sediments that are occasionally formed when lightning discharges into ground; that is, they are not extraterritorial. When this meteorite was first studied in Cardiff it was found to clearly contain embedded microbial structures, while oxygen isotope studies clearly established it as of non-terrestrial origin.

For many years, space scientists have used oxygen isotope studies to determine the non-terrestrial origin of claimed meteorites. The proportion of the stable isotopes of this element are different in terrestrial and extra-terrestrial samples. So, by determining the so-called isotope fractionation one can work out if a proposed meteorite sample is authentic. In order to clear up the matter, Jamie Wallis, one of CW's research students, sent a sample of the presumptive Polonnaruwa meteorite to the Isotope laboratory at Gottingen University in Germany. The results were completely decisive, the Polonnaruwa sample is not from Earth, but is truly extra-terrestrial; the same result came back when the sample was tested in Japan. We can be confident then, that despite what uninformed critics say, the Polonnaruwa meteorite is authentic and we can similarly be confident therefore that any fossilized material it contains came from outer space (Fig. 10.7).

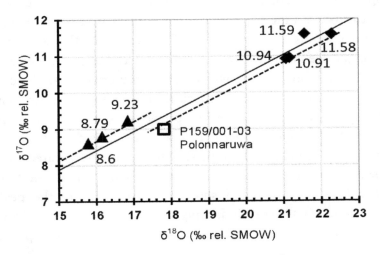

Fig. 10.7. Oxygen isotopic composition data.

The above data is taken from a sample of the Polonnaruwa meteorite (sample P159/001-04), compared with data from other CI chondrite and CI-like chondrite meteorite samples (Alais (8.6), Ivuna (9.23) and Orgueil (8.79), (Meta-C) B-7904 (10.91), Y-82162 (11.59), Y-86720 (11.58) and Y-86789 (10.94)).

Further Details of the Evidence Showing that the Pollonaruwa Stone is Meteorite

Technical details, modified from Jamie Wallis and Richard Hoover's publications, showing that the Polonnaruwa stone is a meteorite are shown below. You can skip these if they are not of interest; you can rest assured however, that the data confirms that a true meteorite is being studied. Further samples of the Polonnaruwa meteorite are fortunately available for scientists to study in the future.

Jamie Wallis reported on the physical, chemical and mineral properties of a series of stone fragments recovered from the North Central Province of Sri Lanka following a witnessed fireball event on 29 December, 2012. He comments: The stones exhibit highly porous poikilitic textures comprising of isotropic silica-rich/plagioclase-like hosts. Inclusions range in size and shape from mm-sized to smaller subangular grains frequently more fractured than the surrounding host and include ilmenite, olivine (fayalitic), quartz and accessory zircon. Bulk mineral compositions include accessory cristobalite, hercynite, anorthite, wuestite, albite, anorthoclase and the high pressure olivine polymorph wadsleyite, suggesting previous endurance of a shock pressure of ~20 GPa. Further evidence of shock is confirmed by the conversion of all plagioclase to maskelynite. Here the infrared absorption spectra in the region 580 cm^{-1} to 380 cm^{-1} due to the Si-O-Si or Si-O-Al absorption band shows a partial shift in the peak at 380 cm^{-1} towards 480 cm^{-1} indicating an intermediate position between crystalline and amorphous phase. Host matrix chemical compositions vary between samples, but all are rich in SiO_2. Silica-rich melts display a heterogeneous K-enrichment comparable to that reported in a range of nonterrestrial material from rare iron meteorites to LL chondritic breccias and Lunar granites. Bulk chemical compositions of plagioclase-like samples are comparable to reported data e.g. Miller Ranger 05035 (Lunar), while Si-rich samples accord well with mafic and felsic glasses reported in NWA 1664 (Howardite) as well as data for fusion crust present in a variety of meteoritic samples. Triple oxygen isotope results show $\Delta^{17}O = .0.335$ with $\delta^{18}O$ (‰ rel. SMOW) values of 17.816 ± 0.100 and compare well with those of known CI chondrites and are within the range of CI-like (Meta-C) chondrites. Rare earth elemental abundances show a profound Europium anomaly of between 0.7 and 0.9 ppm while CI normalized REE patterns accord well with those of high potassium and high aluminium glasses found in lunar and 4 Vesta samples. Two-element discrimination maps of FeO vs SiO_2, FeO vs TiO_2, FeO vs Al_2O_3 and FeO vs Na_2O similarly match those of impact glasses present in lunar samples and remain within relatively close proximity of the KREEP component. Iridium levels of between 1–7 ppm, approximately 10^4 times that of terrestrial crustal rocks, were detected in all samples.

Richard Hoover's evidence

In a PhD Thesis, Dr. Jamie Wallis showed that the Polonnaruwa stone exhibit nonterrestrial oxygen isotopes and physical properties which are different from all known meteorites and terrestrial rocks. **While they have properties which are not seen in all known groups of meteorites, these stones are also substantially different from all known Earth rocks.** No established meteorites have density ~0.6 to 0.8 gm/cm^3 and only two Earth rocks (pumice and diatomite) are able to float in water. However, recent observations show that Comet 67/P and the dark boulders on Near Earth Asteroid Bennu also have densities <1 gm/cm^3. The physical properties and ENAA clearly demonstrate the dramatic differences in density, porosity and element composition of the Polonnaruwa stone as compared to major Carbonaceous Chondrite, Apollo 14 KREEP basalt, Continental crust rocks and Pacific micronodules. The density and porosity of the Polonnaruwa stones is similar only to that of the Comet 67P/Churyumov-Gerasimenko and the dark boulders of the Near Earth Asteroid 101855 Bennu. Clearly, the Polonnaruwa stones are very different from the measured representatives of major groups of carbonaceous meteorites. The Rhenium excess is particularly interesting. Rhenium is a heavy, third-row transition metal in group 7 of the Academia Journal of Scientific Research periodic table with atomic number 75. It is one of the rarest elements in the Earth's crust with an estimated average concentration of one part per billion (ppb). The unusual concentrations of Rhenium, REE's and HPE's to provide additional support to the earlier demonstrated Oxygen Isotope and mineralogy data establishing that the Polonnaruwa stones are not terrestrial rocks. Although the Th/U ratios of the Polonnaruwa stone is not markedly different from the other carbonaceous chondrites studied, their excess of K is astonishingly high.

Neutron activation analysis studies help resolve the mystery of the very low-density Polonnaruwa stone and have revealed very dramatic differences in the density, porosity and element composition of the Polonnaruwa stone as compared with CI, CM, CO, CK and CV meteorites, zircons, pacific micro-nodules and rocks from the continental crust of Earth and Moon. Some of the Lanthanide rare earth elements (REE) are particularly higher (La, Ce, Nd, Sm) while the Europium level is lower than that of other carbonaceous meteorites. The Polonnaruwa stone exhibits astonishingly high levels of the group one alkali metal Rubidium and the group 7 transition metal Rhenium, which have radioisotopes with extremely long half-lives (>40 Ga); Rhenium is one of the rarest elements in the Earth's crust with an average concentration <1 gm/cm^3.

The Sheffield-Examination of the Polonnaruwa Meteorite for Fossils

The provenience of the Polonnaruwa meteorite we were provided with was checked by colleagues (including CW) on visits to Sri Lanka, so we can be certain that it fell to Earth (with many other samples) and was collected off the ground by a local farmer. A sample of the meteorite was forwarded to Sheffield, where it was sterilized with bleach (10% w/v), washed with sterile distilled water and then dried (37°C). The indomitable Chris Rose and Alex Baker then sectioned and examined it under the scanning electron microscope; where appropriate, the samples were analysed using a scanning electron microscope with an attached element analyser (EDAX). (Further details of the methods employed can be found in the Appendix).

What we found within the Polonnaruwa stone is truly amazing and you can imagine how astonished we were when we saw the images which follow.

Figure 10.8 shows clear images of diatom frustules. It is important to note that these are images of inside the meteorite, not on its surface. The diatoms are partially integrated within the matrix material and therefore do not result from ingress of modern frustules. Note for example, how in the top right-hand image, the notch in the frustule can be seen to fit exactly into a projection from the body of the meteorite; even clearer evidence of such integration can be seen by referring to the following image (Fig. 10.9).

The obvious response to these images is that they are modern frustules which have somehow fallen on the cut surface of the meteorite or have gained ingress when the meteorite was possibly submerged in water. The likelihood of the former is negligible and is dismissed by the clear evidence of integration, while we conclude that it is impossible for them to have passed through the small pores in the matrix to reach the interior of the meteorite. The large erect frustule is twenty-five microns in length while the background pores are far smaller.

Fig. 10.8. Images of diatom frustules **located within** the Polonnaruwa meteorite (not an exposed surface). (Bar represents 25 μm)

Fig. 10.9. A Polonnaruwa-diatom frustule which was imaged in Cardiff, showing clear integration of the left side of the frustule with the meteorite-matrix.

The argument that the diatoms we see in the Pollonaruwa stone originated when water seeped into its porous body does not square up; simply because if this were the case then we would also find, in the centre of the P meteorite, all manner of aquatic life, including diatoms, free-living algae, rotifers and terrestrial worms; instead all we find are diatoms. Remember the pond-water paramecia I used to enjoy seeing as a boy with my *Bijou Microscope* — where are they? Why are they not found inside the Polonnaruwa meteorite if it was contaminated with terrestrial water? Diatoms, being rigid, cannot move in and out of small pore openings in the same way as other pond life. If it is claimed that terrestrial diatoms can contaminate the centre of the Polonnaruwa stone, why are smaller and more flexible aquatic microbes and animals absent? The absence of terrestrial, aquatic life in the Pollonaruwa meteorite demonstrates that the diatoms which are present are fossilized and came from elsewhere in the cosmos and not from terrestrial soil, mud and water. Once again, we provide evidence that microbial life is not restricted to this planet.

An internet critic claimed that the frustules shown in the original report of the presence of diatoms in the Polonnaruwa meteorite are too pristine to have made an interstellar journey. It is important to note however, that diatom frustules are extremely hardy. They have for example, been found in 2–40-million-year-old marine Eocene deposits. Frustules also survive intact during the production of filter material, such as Chromosorb which is diatomite crushed, blended, pressed into brick and fired at 900°C and then ground. Diatom frustules have also been shown to partially survive heating to 1100°C. It would seem likely therefore that diatom frustules would readily survive the rigours of a cosmic journey when embedded within a meteorite.

A major objection to a space origin for these frustules is that they are identical to modern diatom. Surely, it is argued, they would be different from modern terrestrial specimens unless, in the unlikely event, they had somehow co-evolved in space. This objection can be answered by again pointing out that modern diatoms are identical to those isolated from Eocene sediments and appear not to have dramatically changed as they evolved during the interim vast time period. It is possible therefore that diatoms have not evolved, but were delivered to Earth aeons ago and have not fundamentally changed since, and of course, in our view they are still arriving on Earth from space.

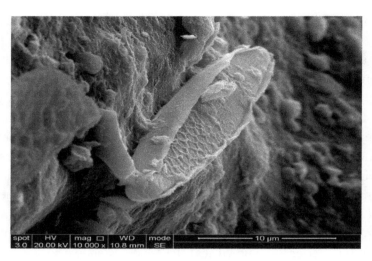

Fig. 10.10. A diatom frustule-half in the Polonnaruwa meteorite, covered in a net-like material which looks like the plasmodial phase of a terrestrial slime mould.

An example of how difficult it is to interpret scanning electron microscope images of the interior of meteorites is shown in Fig. 10.10. Here, we clearly see a boat-shaped half of a diatom frustule inside the Polonnaruwa meteorite. Again, it seems impossible for this large structure to have been washed into the interior of the stone, so we conclude that it is not a terrestrial, modern diatom, but arrived within the meteorite. The diatom frustule is covered in a slime-like material, showing a characteristic net-like appearance.

Microbiologists in the know would likely recognize this as the plasmodial stage of a slime mould. It would make a pleasing story to assume that this net was fossilized and came to Earth from space, via panspermia. Slime moulds have in the main been around on Earth for some 100 million years, while some marine species have been here for over two billion years, without undergoing much evolutionary change. They have the amazing trick of shutting down their metabolism in a process called cryptobiosis. Perhaps slime moulds exist elsewhere in the cosmos, using growth strategies which enable them to survive long periods in space. However, unlike the diatom frustule, which they cover, the slime mould seen in this picture could readily have seeped into the meteorite as it lay in mud and water prior to it being recovered and analysed. The EDAX results however, show that the mineral composition of the slime and meteorite matrix are essentially identical showing that the slime is in fact fossilized.

While there can be arguments about the provenance of diatoms in the Polonnaruwa meteorite, there can be no argument about the biological nature of the object shown in Fig. 10.11. Here we see an oval-shaped structure within the body of the meteorite. The oval shaped void contained within the fossil is surrounded by a double-layered structure with cross linkages suggestive of a plant cell wall (a), (b) parent material and (c) a plant-like palisade layer. EDAX

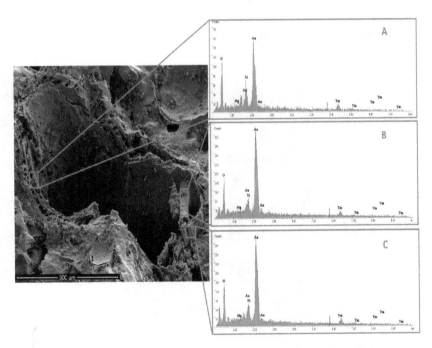

Fig. 10.11. An oval-shaped fossilised biological inclusion and EDAX graph for a complex structure found in the Polonnaruwa meteorite. Note that this image is taken from a cut section.

analysis shows that all of the indicated regions have essentially identical composition and are well-integrated in the body of the meteorite. The matrix matter of the meteorite and the fossilised palisade layers are identical, showing that the palisade layer is not derived from modern plant material that has somehow become embedded in the fossil matrix.

Figure 10.12 shows a detail of an image of a present-day grass-shard (a) and the fossil plant wall found in the Polonnaruwa meteorite; the similarities are obvious. Although we do not necessarily claim that the meteorite-fossil is a grass shard (its oval structure, for example, might suggest a seed), it is certainly plant-like; it could equally be an unknown biological entity of cosmic provenance.

The results show that fossils of complex life, including complex eukaryotes exist in the cosmos and that their fossils are delivered to Earth from space; the obvious corollary being that life is, or was, at some time present elsewhere in the Universe.

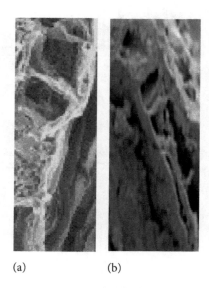

(a)   (b)

Fig. 10.12. (a) Electron microscope image detail of part of the palisade layer of the wall of a non-fossilized present-day terrestrial grass shard and (b) part of the palisade layer of wall of the fossilised biological inclusion.

Clearly, our conclusions, which are based on experimental evidence are likely to prove difficult to accept to a vast body of conservative scientists. The presence of such complex biology in claimed meteorite would be used to stridently maintain that the Polonnaruwa stone is not a meteorite, since by definition (or consensus), meteorites have not been shown to contain evidence for life! The analysis of the Sri Lankan (Polonnaruwa) meteorite samples conducted by Jamie Wallis and Daryl Wallis in Cardiff and independently by Richard Hoover have produced results that are so amazing that they continue to provoke intense argument and controversy. The Sri Lankan meteorite contains within it fossilised *extinct* microbes and fossilised diatoms that are beautifully preserved. These new discoveries clearly confirm that life is a truly cosmic phenomenon. The only way out for the critic is to assert that the "meteorite" sample studied here is not a genuine meteorite, but is a rock from the Earth. Not only does this assertion fall foul of the well-attested fireball sightings that preceded the fall, but it also contradicts the discovery that the meteorite contains a ratio of oxygen isotopes that are not consistent with Earth rocks. Strident rejections that have been voiced in some circles in the West have distinct echoes of racist slurs. In one reported attack (on the internet) it has been stated that Sri Lankans would not have been able to distinguish a meteorite from dried cow shit! But returning to hard science we must mention in passing that Jamie and Daryl Wallis discovered that the Polonnaruwa meteorite contains so high a content of the element iridium that it *cannot* possibly have come from Earth.

Some critics will probably say, yet again, that we are suffering from pareidolia when interpreting plant-like fossil images, a criticism which we have no additional analytical evidence to refute. This argument is not, however, tenable in relation to the observed frustules which are irrefutably diatoms. While it could be suggested that the diatom frustules are terrestrial and somehow washed into the centre of the stone (despite their large size relative to the pore size of the bolide), we yet again emphasize that the same argument cannot be used in relation to the plant-like fossil which is clearly integrated into the body of the rock. Nor obviously could this region have fossilised between the bolide being observed and it being collected and studied. A terrestrial origin for the fossilized biology described here could possibly be explained by the suggestion made by Urey, who concluded that the most probable origin of supposed biology in the Orgueil meteorite is that a nearby planetary body, at some point, became temporarily contaminated with water and life forms from Earth — these having being preserved and are now returning. Whether the introduction of such a currently untestable proposition helps in the matter is, however, debatable. Most scientists will, despite the evidence presented here, likely fall back on parsimony and assert that the Polonnaruwa bolide is not a meteorite and that our findings must somehow relate to terrestrial fossils present in a tektite, or other geological sample (as yet, however, we have failed to find any evidence suggesting the presence of biology in tektites). Most biologists, who are immersed in an Earth-centred evolution-theory paradigm, will find it hard to accept the conclusions we have reached, particularly in regard to our claims that a meteorite contains a plant-fossil, a finding which we claim demonstrates the existence of non-terrestrial life closely similar to life that has evolved on Earth. All we can assert is that we and others amongst our colleagues and collaborators have proved that the Polonnaruwa stone is indeed a meteorite, and this meteorite contains diatoms and fossilised plant life. This work has been peer reviewed and has either been published or accepted for publication by an international scientific journal. Such findings prove that life exist elsewhere in the Universe! There are times when we ourselves can hardly believe what we have found!

Fossilized Worm-like Structures Complete the Polonnaruwa Meteorite Story

Below, (Fig. 10.13) is yet another remarkable photograph of the inside of the Polonnaruwa meteorite. Here, we see in the meteorite matrix

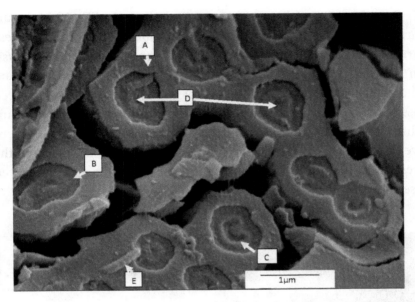

Fig. 10.13. Worm-like structures inside comma-shaped cavities of the Polonnaruwa meteorite.

(a) slightly ovoid cavities, mostly single, (b), but one as a pair. At the base of the cavity we see a comma-shape indentation (c) which in some case contain worm-like structures (d), while others are empty of the worms as shown in more detail (e), having apparently been removed in some way from one of the comma shaped depressions. Since these cavities occur in the cut surface of the meteorite, we can see that when the upper half of the meteorite is put in place, sealed "tomb-like" structure will be reformed; these containing the comma-shaped cavities which in turn contain the worms-like structures. The fact that the worm-like structures, exactly fit the comma-like cavities, and are entombed, shows that they are not contaminating-contemporary organisms. Interestingly, an easily overlooked structure, which is clearly a diatom (and similar to the ones shown above), can be seen in the top left-hand corner of the image.

A mass of worm-like structures can be seen in Fig. 10.14 on the surface of group of cavities. The congregation of a number of these forms provided the opportunity to perform an elemental analysis on them. As can be seen, the worm-like mass contains oxygen and silicon, but no carbon. This shows that worm-like features are indeed fossilized biological structures and not contemporary organisms which have somehow gained entry to the middle of the meteorite, or overlain its surface after cutting. These images again demonstrate the presence of biology in a meteorite confirming that life exits, or existed elsewhere in the cosmos.

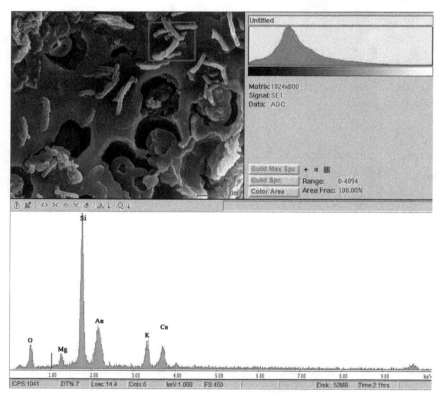

Fig. 10.14. Mass of worm-like fossils and corresponding EDAX, showing them not to be organic, i.e., not contemporary living bacteria.

11

Proving that Microbial Life of Space Origin Exists in the Stratosphere

Some Personal Recollections (MW)

Beginning sometime in the mid-1980s one of us (MW) began to teach panspermia to around one hundred and fifty first-year Biology students, taking, a course on general microbiology. The lecture was given to provide an alternative to the then, as indeed now, generally accepted theory that life began when chemicals in a primordial "soup" came together from first amino acids and the proteins (more of which later). At that time panspermia was largely ignored or rejected by the scientific establishment and few undergraduate students anywhere in the world would have been exposed to the idea. It certainly was not to be found described in detail in run of the mill first year biology texts. The idea that MW was teaching even a single lecture on the subject was frowned upon by the majority of his departmental colleagues, notably members of the teaching committee, a situation that remained until MW retired some thirty years later. For the first time it became amply clear that while the established academics shunned panspermia, students in contrast, found the subject fascinating.

When reading up on the subject MW, of course, could not fail to become acquainted with the name of co-author, Cardiff astronomer and mathematician, Chandra Wickramasinghe. Panspermia is a conjoint subject made up of astronomy and biology — notably microbiology, since it involves the claimed transfer of microbial life across the cosmos. Not surprisingly, MW was keen to meet CW, and even collaborating with him in some way. MW assumed that since CW was head of an academic department in Cardiff (he was Professor of Applied Mathematics and Astronomy) that he would have eminent microbiologists on his door step, to advise him; alternatively, if they were not interested, then microbiologists elsewhere in the country, or even the wider world, would be falling over themselves to become involved. It was with such thoughts that co-author MW, kicked the idea of a possible collaboration into the long grass and got on with his own, firmly Earth-based, research on environmental microbiology, involving prosaic subjects like sewage and soil pollution and soil biochemistry. Fate then intervened, as it seemed to do throughout the development of our story. As MW was a member of the UK Society for Microbiology, he received the Society's publications, including one that published letters of general interest from members. One such published letter that enraged MW was basically a rant against panspermia in general and CW in particular. MW promptly wrote and published a riposte to the letter and assumed that was the end of the matter.

As fate decreed, however, co-author CW read this letter and promptly contacted MW immediately enquiring if he might be interested in discussing panspermia with a view to collaborating on a joint publication in his capacity as a microbiologist. While a few microbiologists at Cardiff seemed happy or comfortable at the time to work with CW, for political reasons (which were to always interact with this story) he welcomed the prospect of a microbiologist from outside Cardiff to join his team.

"Team" is something of exaggeration. Chandra Wickramasinghe never had a team in the sense that there were scientists in laboratories in his department who were working together on panspermia. Instead he had to find collaborators wherever he could, anyone who was happy to suffer the opprobrium of their peers and work on panspermia. Some, would provide access to scientific equipment not housed in the Cardiff Astronomy or Mathematics Departments, like electron microscopes, while others would have the ability to launch probes into the stratosphere. Other collaborators were purely theoretical aides adding their knowledge of astronomy and space science to the mix. Several unsuccessful attempts were made at this time to get funding from the UK research councils for this Work, but of course it was clear that committees do not fund heterodoxy in science. In our later collaborations CW and MW had to grub around for funding where we could, using all kinds of "creative" strategies. This work was often funded by private individuals without any allegiance to institutions, whose only interest was in finding the answer as to how life originated. We certainly were always severely restricted in what we could do on account of the lack of funds. But as Rutherford posted on a sign in his laboratory — "We have no funds, so we must use our brains". Having restricted funding is of course very frustrating, but we can understand Rutherford's message from our own experience. As we shall see, by lacking funds, we were redirected into more innovative approaches which we would never have become involved with otherwise. Of course, the opposite is also true and we could have (and could now do) a lot more had we been flush with funds. But it is what it is, or rather was what it was!

At the first meeting between the present authors CW got straight to the point and asked of MW if he would try and isolate microbes from some samples he already had in his possession. These had been collected from 41 km in the near space environment of the stratosphere with Indian collaborators, including the celebrated cosmologist Jayant Vishnu Narlikar under the auspices of the Indian Space Research Organisation (ISRO).

MW would eventually travel to India and become involved in this stratosphere-sampling work, but for the moment, he was content to take the samples provided by CW back to Sheffield and try and isolate anything that could be found in them. Isolating microbes was then becoming a lost art as most microbiologists were beginning to rely upon DNA methods to identify microbes directly in the environment. Growing bacteria and fungi is a bit like gardening, an art form. Fortunately, MW received his apprentice training as a microbiologist before the advent of DNA-based methods and so was well acquainted with the "old-fashioned" media-based methods, for which you need the equivalent of "green fingers". These methods involve the use of agar contained in plastic petri dishes (when MW first started his career these were made of heavy glass). Specially selected nutrients are then added to the agar before it is poured (when the mix becomes referred to as a medium), in order to specifically grow individual organisms. Fungi (i.e. moulds) for example, tend to prefer sucrose as a source of carbon foodstuff, while bacteria often selectively grow on glucose. Some bacteria, such as *Helicobacterium pylori*, the cause of stomach ulcers are what is termed "fastidious" and need to be provided with a complex medium in order to grow. Even before the attempt at isolating microbes from the stratosphere samples had started it was expected that it would be hard work getting anything to grow. For one thing the samples would have been exposed to the extremely low temperature conditions and UV-radiation exposure found in the stratosphere and so would likely not grow on a sugar-rich medium at laboratory incubation temperatures. This likelihood was empathised by the fact that, despite numerous attempts, none of the microbiologists were able to isolate anything from the Indian samples. But amazingly, MW soon got something to grow, not because of any genius as an isolation-wizard, but because of sheer luck, or serendipity. Serendipity is a word is of Arabic origin, first used in AD 361. It relates to the old word, Serendip, for the island of Sri Lanka, formerly Ceylon. The word was coined by Horace Walpole based on the inspiration of a Persian fairy tale," The Princes of Serendip", whose heroes made discoveries by chance. The most famous example of serendipity is of course the discovery of penicillin by Sir Alexander Fleming. Other examples include pulsars, radioactivity, vulcanised rubber and x-rays. In fact, serendipity is such a frequent and important means of scientific discovery, that MW used to suggest to his students that they actually plan for it; for example, by leaving, exposed to the air, samples of any biological experiment they were working on, and let microbial contamination enter the frame. Of course you have to be an experimentalist to be exposed to the benefits provided by serendipity and you have to spot what it sends you, or as Louis Pasteur famously said, "Fate (serendipity) favours the prepared mind". There is no way that Donald Trump, for example could have discovered penicillin by some happy accident! We will see later that serendipity once again comes into this story and dramatically changes the course of our research.

Back to the role of serendipity in relation to our stratospheric research program. One of the benefits of having PhD students is that one can get them to do the more mundane jobs, leaving you, the experienced scientist, to concentrate on the more enjoyable, fun — specialist and innovative work (university technicians who were employed in the past are, it turns out, a dying breed). So MW being a minor *prima donna*, asked one of his PhD students to prepare a bacterial growth medium which he would then use to try and isolate bacteria from the stratospheric samples. The expectation was that he would find a clear, pristine medium in the petri dishes left by the student in the area of the laboratory devoted to sterile, microbiological work. Instead, he found a collection of petri dishes containing a brown murky medium. This was surprising, as well as annoying from, the outset!

The student could have started again and provided MW with proper clear samples of the medium, but as it was the end of the day MW decided to use the ones in front of him, without delay. So using sterile techniques he transferred some of the stratospheric sample into the centre of one of the murky plates and then transferred the petri dish to a 37°C laboratory incubator and left it incubating overnight. On inspecting the plates, the following day, amazingly characteristic, slimy growth of a bacterium could be seen around the sample. Next, some of this slime was transferred, using long-practised sterile techniques, to a fresh dish and re-incubated. This was done to ensure that we had a sample of the isolated bacterial culture, just in case something untoward happened to the original. Then, MW sent off the newly isolated bacterium for DNA analysis by an independent laboratory. This would take a heavy toll on MW's meagre research funds, but it was realised that an independent analysis would hold more sway than one conducted in "house".

A week or so later, the result arrived and the email result was read with excitement. Two bacteria had been unequivocally isolated- *Bacillus simplex* and *Bacillus pasteuri*; subsequently the same strato-spheric sample yielded a fungus, which was independently identified as *Endyogontium album*. MW and colleagues, then published a scientific paper on the successful isolation and so informed the world that bacteria and fungi can exist in the stratosphere at a height of 41 km (around 25 miles).

Some years later a balloon flight launched to a height of 41 km in 2008 led to the recovery of more stratospheric material, and analysis by Dr. S. Shivaji and his colleagues in India yielded cultures of three hitherto unknown microbial species which were all highly resistant to ultraviolet light. One of the newly discovered species was named *Janibacter hoylei*, in honour of Sir Fred Hoyle. All the new bacteria that were discovered had a large fraction (80%) of their DNA identical to terrestrially common phenotypes, but they were sufficiently different to be listed as "new" species.

The work we have described, however, did not mark the first claimed recoveries of bacteria from the stratosphere. As early as 1970, Russian scientist, Imshenetsky reported the isolation of microbes at a height of 61km in the stratosphere. Nevertheless, ours involved the use of more modern sampling and more reliable microbial analysis techniques.

So where did serendipity enter our own story? To recap, we have mentioned that we asked the PhD student who had obtained a darker than usual bacterial medium by "mistake". It was from this sample that MW had attempted the isolation of microbes from the original stratosphere samples. In order to save researchers time many media are made up by chemical companies and sold to scientists in bottles.

Like similarly sold food products these bottles are stamped with a use-by-date. On this occasion, not being experienced, the student had used a very old bottle of medium, which should have long-since been discarded. The resultant growth medium failed to set properly and so by accident, the student gave me what is called a "soft medium". One of us (MW) immediately recognised this but still decided to use it. Now, any microbes present in the stratosphere would likely be freeze-dried as the result of the conditions present there. Freeze drying is used to store representative samples of bacteria in culture collections; in this state they remain alive, but dormant, allowing them to survive for very long periods, thus enabling them to be forwarded to research workers who then resuscitate the bacteria. And the recommended way of resuscitating such viable freeze-dried bacteria is — you guessed it — to use a soft culture medium. Thus by pure accident the student having prepared a soft medium had provided the best conditions for reviving freeze-dried bacteria from an exceptionally cold environment like the stratosphere; by not throwing this accidental medium away to employ instead normal hard medium, serendipity played a crucial role. So, had the student known what he was doing, we would probably not have isolated bacteria from the stratosphere on this occasion.

Both of the present authors were obviously excited first by this discovery and subsequently in having the work published in a mainstream microbiology journal. Then we were shocked by a bolt from the blue! A very senior biologist at Cardiff University informed the then Vice Chancellor that our work resulted from contamination and that we were somehow cheating, simply put, we were accused of being lying charlatans. CW was hauled over the coals, and of course was determined to defend our position and reputation. There is no doubt that some of the Cardiff scientists were surprised, shocked, and doubtless disappointed that we isolated microbes from the stratosphere samples so quickly, but there was absolutely no reason for them to attack us with such malicious claims.

Two overseas research trips relating to the question of microbes in the stratosphere stand out. One was a visit that MW made to the

Lawrence Livermore Laboratory in California. This is a laboratory famous for its work on physics and its association with the Star Wars Initiative. MW had the opportunity to visit, and with the help of the physicists, P.K. Weber, J.B. Smith, and I.D. Hutcheon, to use a complex piece of equipment, called a nano-sims. One way by which we can determine if bacteria, isolated from the stratosphere, come from space is to determine their carbon isotope ratios. Carbon has two isotopes, C14 and C12. Earth-derived organisms are composed almost exclusively of the C12 isotope, while it is expected that space–derived organisms would contain a significantly different mixture of the two. We can theoretically determine the isotope composition of microscopic living organisms using a nano-sims machine connected to a scanning electron microscope. The visit to the Lawrence Livermore was in the hope of doing precisely this. MW was accompanied by Brig Klyce, the author of a first-rate website on panspermia, called *Cosmic Ancestry*. Although not trained as a scientist, Brigg had become extremely knowledgeable about this topic. Unfortunately, our visit proved to be fruitless. The nano-sims we found at Lawrence Livermore was an amazing, largely home-built machine, but it possessed a major weakness from our perspective, namely it was equipped with a relatively poor scanning electron microscope. While this component of the set up was fine for the work done by physicists we were unable to locate our extremely small bacteria with it and therefore could not determine their isotope ratios. Yet the problem remains how did they get to this region?

There can be no doubt then that microbes exist in the stratosphere at heights of up to 41 km and probably far beyond. The question is — are they incoming from space, or are they as most scientists tend to suggest "lofted" from Earth to this extreme height?

12

Evidence that Life is Continually Arriving from Space — Neopanspermia

We have already referred to the results of a few investigations carried out by co-author MW on samples of stratospheric material supplied to him by CW.

These were obtained from a balloon-sampling experiment carried out over India (Fig. 12.1). A very large balloon was used to carry what is called a cryosampler into the stratosphere at a height of up to 42 km (around 25 miles). You can gauge this height by thinking of the flight you made in a holiday jet which generally cruise at around 10 km (Further details of the methods employed can be found in the Appendix).

The balloon carried evacuated cryosampler-tubes cooled *in situ* to liquid He temperatures; large amounts of atmosphere were then sucked when seals on the tubes were opened remotely as the balloon crossed the stratosphere. These metal tubes were then parachuted to Earth and washed out with buffer in the laboratory under strictly controlled aseptic conditions. The washout was passed through a filter-membrane with extremely small pores, in order to trap and isolate any microscopic structures that can be captured in the stratosphere; any living microbes present on these can then be grown (Fig. 12.2).

Additionally, the surface of the filters was scanned using a scanning electron microscope in order to examine any organism at extremely high magnification. We also used a sophisticated ultraviolet light microscope to see any living objects, exposed to so-called "vital stains", which amongst other things, can locate the presence of DNA (Fig. 12.3).

Because the entry point into the cylinders was restricted in size, no large microorganisms such as algae and fungal spores, pollen grains

Figure 12.1. Left to right: the very large balloon used to lift the cryosampler (shown here in its transport cradle), and Chandra Wickramasinghe holding one of the cryosampler tubes.

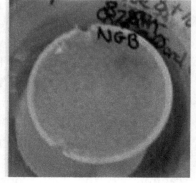

Fig. 12.2. New, pristine white micropore-filters before filtration of cryotube washout and, after filtration, covered with cosmic dust.

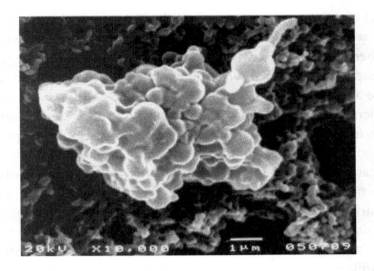

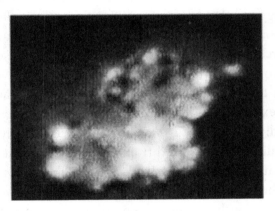

Fig. 12.3. Large bacterial mass isolated from 41 km and the same mass stained with carbocyanine stain.

and grass shards were allowed into the sampler. As we have seen, in 2003, we reported the isolation of two bacteria (*Bacillus simplex* and *Staphylococcus pasteuri*) and the fungus (*Engyodontium album*) from a single stratosphere sample obtained from the 2001 balloon flight. In a separate study, two further bacteria (*B. licheniformis* and *B. pumilus*) were isolated (in India, by Shivaji and colleagues) from a different subset of the same stratospheric samples. As a result of the first ISRO 2001 stratospheric collection project, it was possible to estimate the number of bacterial cells collected in a measured volume of the stratosphere that entered each cylinder from the height of 41 km. This led to an estimated in-fall rate over the whole Earth of 0.3–3 tonnes of microbes per day. Such an infall converts to some ***20–200 million bacteria per square metre arriving from space every single day.*** This truly vast number unfortunately pales into insignificance when compared to bacteria and viruses originating from the Earth's surface, some of which could be lofted to heights of about 3 km in upward air currents and brought down in mist and rain. The average flux of mainly recycled viruses was found to be ***800 million per square metre per day and an estimated number of bacteria amounting to 10 million per square metre per day.*** The close similarity of these estimates suggests that a significant fraction of the flux discovered may have come from the stratosphere and ultimately from space.

A research student of MW, Tariq Omari also used molecular techniques to demonstrate the presence of so-called extremophile bacteria at a height of around 30 km; bacteria which are adapted to growing in extreme environments on Earth. The following bacteria were isolated: *Oceanobacillus iheyensis* — An extremely halotolerant and alkaliphilic species isolated from deep sea sediment at a depth of over a thousand meters on the Iheya Ridge; *Prochlorococcus marinus,* A photosynthetic marine extremophile; *Flexistipes sinusarabici.* A type of bacteria previously isolated from Atlantis II Deep brines in the Red Sea; *Persephonella marina,* A bacterium previously isolated from deep hydrothermal vents and adapted to growth at high temperatures; *Salinispora tropica,* An obligate marine actinomycete; *Kineococcus radiotolerans,* A radiation resistant microbe, previously isolated from a heavily irradiated area of the Savannah River, a nuclear reservation in South Carolina, and finally *Marinithermus hydrothermalis.* The latter which, as its name suggests, is a thermophilic marine bacterium was previously isolated from a 1,300 metre deep-sea hydrothermal vent chimney in Japanese waters. Tariq also isolated *Aquicella siphonis,* a species of bacteria which has been found growing in the Lascaux Cave, of cave-paintings fame. None of these bacteria are commonly found in the laboratory environment, so their isolation does not result from laboratory contamination.

In further DNA-based studies, Tariq showed that a vast array of bacterial DNA is present at around 24–280 km in the stratosphere. He is also confident that he isolated human DNA from this region; although this claim can obviously be challenged on the grounds of likely contamination. This finding however, if corroborated, opens up the intriguing possibility that small particles of human material such as sub-micron skin-fragments could be continually elevated from Earth's surface to the stratosphere, and carried across the globe in stratospheric air currents, and then possibly transferred into interplanetary space.

A most intriguing finding in relation to our own work was recently reported by Russian workers of bacteria (species of *Delftia* and *Mycobacterium*) found on the outside windows of the International Space Station that orbits the Earth at a height of 400 km. Although these organisms are found on Earth, there is no known mechanism by which they could have been carried to the orbiting height of the ISS. How these bacteria reached the ISS and also how they survived the rigours of the environment at this height remains a mystery. As with our own isolation of bacteria and fungi from the stratosphere (at 41 km), bacteria on the ISS could have originated either from space or the Earth; the mechanism by which bacteria could reach the height of 400 km has yet to be explained. A recent paper by Berera and colleagues notes that vertical winds can reach velocities of up to 250 meters per second, and carry nanometre-sized, and possibly larger bacteria, out into space The probability of such an escape seems extremely low, but one thing we can be certain of that neither this, nor any other mechanism, could elevate the 10–40 plus micron biological entities we find, to the very high stratosphere. Incidentally, since the Russian workers used molecular methods to identify bacteria present in the stratosphere they would never have come across the biological entities (which we describe below) of the kind that we have discovered much lower in the stratosphere.

For some inexplicable reason, NASA has failed to engage with this Russian work, even though it was written up in a peer-reviewed scientific publication; one can only wonder why they have not (as far as we can tell) repeated these Russian studies.

As we said earlier, we have also scanned the surface of some of the membranes through which the ISRO cryotube-stratospheric air was filtered.

Figure 12.4 shows a scanning electron microscope image of what appears to be a clump of bacterial cells. The clump (which we colloquially refer to as the "puppy particle"). The clump is about 10 microns across and is made up of individual bacteria-like cells. Although, the clump has the appearance of being made up of bacteria, experience shows that inorganic particles present in the stratosphere often appear to be microbial cells when they are not. Elemental (EDAX) analysis showed that the entire particle lacked large amounts of mineral elements which are typical of cosmic dust particles, e.g.

Fig. 12.4. A presumptive bacterial clump from the stratosphere collected at a height of 41 km (Bar represents 10 μm).

silicon, iron and heavy metals. As a result, we can regard the clump as being organic in nature and, since the individual particles are of the size of bacteria we are pretty certain we are seeing here a clump of bacteria, isolated from the stratosphere.

How then did this bacterial cell mass reach a height of 41 km? It is generally accepted that no mechanisms exist which could transport a large particle (as opposed to a gas or volatile) from Earth to the highest regions of the stratosphere. While theoretical possibilities have been suggested to account for how a one-micron particle (the general size of an individual bacterium) might be elevated to the lower regions than at present it is impossible to explain how a ten-micron masses like the one shown in Figs. 12.3 and 12.4 could reach a height of 41 km. It might be argued that the clump was formed by the coming together of individual, smaller bacterial cells which had individually reached the stratosphere; alternatively, individual cells could have clumped together in the cryosampler during storage, or during filtration. It is worth noting that any bacteria present in the stratosphere and in the cryosampler will be freeze dried and metabolically inactive and therefore could not grow to form a clump like the one seen here. Perhaps individual bacteria could have clumped together onto the membrane when the membrane was held at room temperature, but again this seems to be highly unlikely.

We see another bacterium in Fig. 12.5 (labelled "b"). This image shows a presumptive bacterial cell (or clump of cells) attached to a large inorganic crystal which EDAX analysis shows to be made up of zirconium (z). The formation of slime-like material (indicated by the arrow) suggests that the attachment is biological in nature. Again, there is no known mechanism by which this zirconium particle, plus its attached bacterium, could have been transferred from Earth to a height of 41 km. This zirconium particle plus bacterium was therefore captured by the crysosampler at a height of 41 km during a rapid descent to Earth. It could be that these components, the bacterium and the zirconium crystal, originated from satellites or space stations or from man-made space debris. Such a possibility would depend on

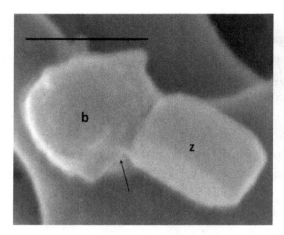

Fig. 12.5. A presumptive bacterial clump (b) attached to a zirconium crystal (z). (Bar represents 0.5 μm).

whether or not these man-made objects are sterilized prior to being transferred to space. Probability enters into the argument; are there sufficient bacteria present in the stratosphere, derived from these objects, to explain the increasing independent reports (based on a variety of sampling methods) of the presence of microbes in the stratosphere. Statistically this seems unlikely considering the vastness of the stratosphere and the fact that the individual sampling events were sent up from widely separated areas on the Earth's surface. We are therefore left with the very high probability that the presumptive bacteria originated from space, and are incoming to Earth.

Since the tropopause is generally regarded as a barrier to the upward movement of particles it is difficult to see how such microorganisms can reach heights above 17 km. Volcanoes provide an obvious means by which this could be achieved, but these occur infrequently and any microorganisms entering the stratosphere from this source will rapidly fall out of the stratosphere. It is difficult to understand how ten micron clumps could be ejected from the Earth to this height, suggesting that such bacterial masses are most plausibly incoming to Earth.

Although there seems to be no obvious means by which micro-organisms could regularly reach 41 km, a number of possibilities can be suggested. For example, it is possible that particles, including microorganisms, are transported to the stratosphere in the updraft resulting from so-called blue lightning strikes. These are conical blue jets that travel upwards from cloud tops and travel at a hundred thousand metres per second to a height of 70 km, charging the atmosphere as they go; whether microorganisms could survive such charging and the high rate of upward movement is of course open to question. Microorganisms might also be carried above the tropopause when thunderstorms and forest fires occur together. Apparently, forest fires strengthen thunderstorms, which then pump immense plumes of smoke and soot (as well as presumably microorganisms) into the stratosphere. Finally, so-called gravito-photophoresis (GP) might provide yet another mechanism by which small microorganisms could reach the stratosphere. As a result of GP, micron-sized soot particles can be carried to a height of 10–20 km, and beyond. Calculated vertical velocities exceed settling velocities by a factor of 30, and it takes about 30 years to transport soot from 10–20 km, and a further 20 years to transport it from 20–80 km. The effect of GP is strongly altitude dependent with the lofting force most effective at an altitude of between 10 and 85 km. It is therefore possible that small microorganisms (less the five microns) could be carried by this mechanism on soot particles. Such transfer would be particularly advantageous, since a coating of soot would protect the organisms from the damaging effects of ultraviolet radiation. However, a major problem with this suggestion is that GP only works on soot particles of diameter around one millimetre; submicrometre-sized bacteria might however, reach the stratosphere by this mechanism, either independently or when travelling on soot particles. This caveat immediately excludes the possibility that fungal spores, which are relatively large, are carried up from the Earth to the stratosphere in this way. It is possible, however, that sub 5-μm microscopic fragments of fungal hyphae can be carried from Earth to this region and then, infrequently, cultured from stratospheric air samples.

We think therefore that the stratospheric microflora is made up of two components: (a) a mixed population of bacteria and fungi derived from Earth, which can occasionally be cultured; and (b) a population made up of clumps of, viable but non-culturable, bacteria which are too large to have originated from Earth; these, we suggest, have arrived in the stratosphere from space. Finally, we speculate on the possibility that the transfer of bacteria from the Earth to the highly mutagenic stratosphere may have played a role in bacterial evolution.

It is particularly noteworthy that bacteria and fungi growing under low nutrient (i.e. oligotrophic) conditions are usually much smaller than equivalent cells grown in nutrient rich environments; oligotrophically growing microorganisms are therefore likely to be the requisite size to enable them to be transported through the stratosphere.

A question which is often asked is — are the stratospheric bacteria and fungi we isolate living or dead? This can be answered by reference to Fig. 12.6.

This shows living bacteria (stained-yellow or green) and dead bacteria (stained red) found on membranes exposed to the stratospheric air (41 km) from the Indian balloon mission. Dead bacteria, staining

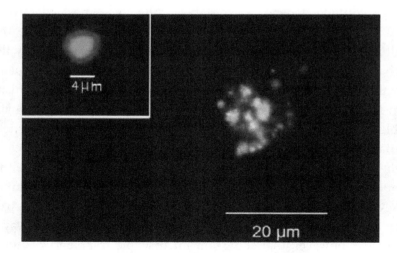

Fig. 12.6. Particles stained using a Live/Dead (BacLight) Stain.

red, were far more numerous on the extended field of view, but this image shows that some bacteria are alive in the stratosphere when captured by the cryosampler.

We can conclude this section by saying that there exist at least two separate populations of stratospheric microorganisms. One population consists of common Earth bacteria and fungi that are carried on a relatively regular basis (by phenomena such as blue lightning, fire-associated storms and GP) to heights above 17 km; these organisms can be cultured, albeit rarely, from stratospheric air samples. The fact that such bacteria are, essentially, genetically identical to the same species derived from Earth suggests to us that they are from Earth, but of course they could have originally come from space. The second component of the stratospheric bacterial population consists of bacteria that exist as relatively large clumps; this population is viable, but non-culturable. It is assumed here that such clumps (around 10 mm across) are too large to have originated from Earth and their distribution in the atmosphere indicates that they have a non-terrestrial origin. Since the majority of bacteria found on Earth have yet to be cultured, the possibility exists that an unknown fraction of these bacteria may have originated from space. The possibility still remains however, that all components of the stratospheric microflora are incoming to Earth. We recognize of course that our views are controversial and will need to be verified by further studies of the microbiology of the stratosphere.

Could the Transfer of Microbes from Earth to the Stratosphere Have Influenced Evolution on Earth?

It has occurred to us that the possibility that transfer of bacteria (and other microorganisms), whether incoming or exiting from Earth to the stratosphere may have impacted on (and continues to impact) bacterial evolution. The continued development of life on Earth has depended upon the presence of an atmosphere that shields the biota from the lethal effects of ultraviolet. Such protection will necessarily have reduced the exposure of bacteria to the potentially mutagenic ultraviolet and other radiation. This problem can obviously be overcome if a mechanism exists whereby bacteria can be transferred from the relatively benign environment of Earth to the highly mutagenic conditions found in the stratosphere. The above described mechanisms provide a means by which such transfer of bacteria between Earth and the stratosphere can occur. Those bacteria that are mutated, but not killed, by exposure to stratospheric ultraviolet radiation (a mixture of UV-A, B and C), will be returned to Earth as the result of sedimentation. The novel, multiple or increased number of mutations they carry with them will enhance the mutation pool that is available for natural selection to act upon and, as a result, the rate of bacterial evolution will have been greater than if such evolution was solely dependent on mutations occurring below the stratosphere. It is likely that such mutations will occur most often in the lower regions of the stratosphere where numbers of bacteria are greatest and where residence times allow for rapid mutation, but not sterilization. Of course, such a view is dependent on a large number of bacteria being exchanged between the Earth and the stratosphere. However, even if we consider volcanic transfer alone, the number of bacteria transferred is likely to be significant over the aeons that bacteria have existed on the Earth.

Bacteria, and other organisms incoming to Earth from Space would have, and will continue to provide new genetic information which, we suggest has always aided the evolution of life on Earth.

Large DNA-Staining Biological Entities in the Stratosphere

We expect DNA and RNA to be the universal biological information and replication system in life throughout the cosmos. An obvious experiment to do therefore is to check to see if what, what we believe are organism in the stratosphere, contain DNA. Melanie Harris obtained the images below (Figs. 12.7 and 12.8) which show amorphous masses around 20–30 µm in size; these stain yellow-green showing the presence of a DNA sensitive stain. We believe that these objects, or masses of smaller objects (possibly bacteria) are too large to have been uplifted from Earth to 41 km and that they are DNA-rich life forms incoming from space to Earth.

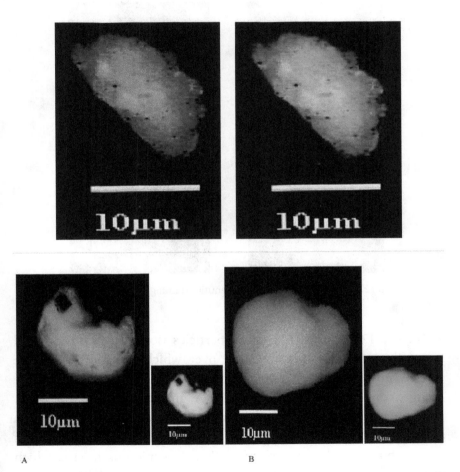

A   B

Fig. 12.7. Large particles from the stratosphere, staining positive for DNA.

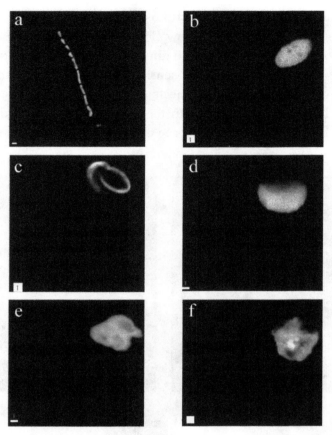

Fig. 12.8. More unusual biological entities staining positive for DNA.

Figure 12.8 shows an array of particles isolated from 41 km, all larger than the five microns limit over which particles cannot be carried into the high stratosphere. Again a DNA stain (acridine orange) confirms that they contain DNA, and are therefore biological. We conclude that it remains difficult to explain how microorganisms are elevated to 41 km and possibly above. Some modelling studies suggest however. that particles less than five microns in size, which include bacteria, and fungal hyphal fragments can **possibly** reach this height. This five micron cut off is extremely important. We base our claim that stratosphere-isolated life forms come from space largely on this point. Stratosphere biology smaller than this could, possibly have been elevated from Earth (or of course could still be incoming from space), but anything larger than five microns must be incoming to Earth from the cosmos, a point which is critical to our argument given in Chapter Thirteen.

13

Strange Biological Entities in the Stratosphere-Personal Recollections of MW

Having confirmed that bacteria and fungi exist in the stratosphere and after publishing the results, we gave a number of local and international lectures, which were generally well received. We were hopeful that funding for continuing this work would follow naturally in view of the immense interest that had been aroused. Then bad news arrived! The Indian scientists asked us to contribute some funds in order to continue having access to their stratosphere-derived samples and actively collaborating with them. This seemed reasonable enough, so we set about trying to obtain grants from UK grant-giving agencies in order to continue our work. We had of course attempted this before, but without success, so we were not surprised when on this occasion as well a contribution to funding was denied.

Fate intervened, however, when MW received an email from a colleague pointing out that a couple of Sheffield University research students were launching balloons in the countryside nearby. MW contacted them and excitedly awaited their visit to his office. The students, Chris Rose and Alex Baker (Fig. 13.1) were immediately recruited and the stratospheric balloon project resumed.

They were of course excited by the possibility of getting involved in the stratosphere research project, and more amazingly, they offered to design and build samplers for the project. People often criticise

Fig. 13.1. Chris Rose and Alex Baker.

today's students for lacking initiative, but these two proved to be amazingly enthusiastic and proactive. At the time they were setting up a business based on launching balloons and hopefully making money out of a variety of popular initiatives, such as offering to photograph birthday greetings or even proposals of marriage against a background of near space, at altitudes just below thirty kilometres. They would later develop and diversify the business into a company called "Sent into Space", and developed stratosphere advertising campaigns. They even appeared on the TV programme, *Dragons Den* with the wonderful idea of offering a service to distribute the ashes of loved-ones into space! Later, they did exciting work with schools, offering children the chance to becoming actively interested in the science of near space.

Within a few weeks, Chris and Alex returned to MW's office with the first design for a stratosphere sampler. We were aware that we could not afford to build a cryosampler of the type we used in India, and so had to go for a much simpler and cheaper alternative; one which could be elevated as high as possible by a single helium-filled weather balloon. By not having funding, fate again intervened and made us completely alter our sampling strategy.

We decide to choose a surface sampler that is, instead of collecting the sparse atmosphere of the stratosphere in a cryotube tube, we went for a sampler, one in which particles present in the stratosphere would alight on the surface of discs which could then be placed in scanning electron microscope; any organisms present could then be viewed and photographed. So we needed a system that could open up in the stratosphere and then close to prevent any contamination, during both the ascent and fall through the heavily contaminated atmosphere. MW suggested using a CD-drawer, which simply opens and closes by remote electric command; it was however, up to Chris and Alex to develop this idea and build the final sampler (Fig. 13.2). Chris and Alex have subsequently designed and built much more elaborate stratosphere samplers, often with the use of 3D printers.

So, a CD-based drawer system is a sampler box and we were ready to conduct the first launch. Amazingly, both Alex and Chris are not only expert balloonists, but their PhD studies involved the routine use of scanning electron microscopes and associated hardware; not only that, they had access to scanners which were far more sophisticated than the ones which we had access to in the Biology Department at the University of Sheffield. Alex and Chris have only a passing knowledge of microbiology, which surprisingly is useful, since they could never photo-shop the convincing biological images they provided; this acted as a kind of built-in control.

Fig. 13.2. CD-Drawer sampler and sampling discs for the scanning electron microscope.

So, just down the road from MW's laboratory then, were two amazingly proactive and enthusiastic students who not only launched balloons, but were expert in the use of scanning electron microscopy electron microscopes and finally, had access to sophisticated versions of this machine. This appears to be so incredibly fortunate that it begs the question of the involvement of Fate (whatever that is) in science; a rationalist, or cynic would of course just dismiss such a suggestion as mere coincidence. It is interesting that Sir Alexander Fleming for one, expressed the same feeling — that destiny somehow intervened when he discovered penicillin. It seemed to him that he was merely the receiver of the contaminated petri dish and thus the motivation to study the phenomenon further.

It would perhaps be useful to discuss here what exactly a scanning electron microscope (EM) is. In the standard EM, extremely fine sections of biological material are exposed to an electron beam allowing us to obtain a highly magnified image inside biological material. As the name suggests, the scanning EM in contrast can magnify extremely small features, or bodies on the surface of a material. Additionally, the scanning EM can be attached to what is called an EDAX machine. This allows the operator to target a chosen spot and determine its elemental composition. By using a scanning electron microscope, we were able to examine the surface of special carbon discs exposed to the stratosphere and look for, and image microbes. By using EDAX, we were then able to determine which materials were made solely of carbon (suggesting a living material) and those containing only inorganic elements, like silicon and iron which indicate non-biological material, such as cosmic dust. We shall see later that this ability to distinguish between living and none-living material in our stratosphere samples was crucial to our work.

The first balloon sampling flights carrying a surface sampler were launched from Chester and subsequently from around Bakewell in Derbyshire. MW was not involved in these initial sampling flights, nor in the electron microscope studies, which were all conducted by Alex and Chris and associated technicians. MW's role was to use his microbiological expertise to interpret the images that were transferred to his computer screen. When the first images arrived, MW was excited beyond belief. To be truthful he had not expected the flights to produce much. Yet here on the screen in front of him were amazing images of what appeared to be microscopic organisms, but even more excitingly, unusual organisms the likes of which he had never seen before in his near fifty years as a biologist (including here my childhood experience looking down a microscope and collecting butterflies and moths, both providers of an awareness of the morphology of organisms and their biology). Because MW did not know (and still does not know) what group of organisms they belong to he referred to them from the start as "Biological Entities" (BEs). They certainly were not typical terrestrial algae, bacteria, fungi or protozoa (i.e. typical microbes) nor were they pollen grains. Of course, the suspicion was that they were not Earth organisms at all, but are instead coming in from space to Earth; there would be a lot of work ahead to prove this hunch.

Subsequently, we obtained just enough funding to conduct further balloon-sampling flights over the UK, as well as Iceland, Death Valley and the US Prairies. Biological entities were isolated from all of these locations, showing that their presence in the stratosphere is not restricted to the UK, but is probably a worldwide phenomenon.

From the beginning, critics again opined the obvious belief that it was more likely that, as our planet is teaming with microbes, that our stratosphere-isolated BEs originated from Earth. If this were the case, then we would expect to find other Earth-derived biology, notably grass shards and pollen grains, mixed in with the BEs found on our sample-discs; in fact, we would expect such normal biology to swamp our discs. Yet we have never found such remnants, nor for that matter, any typical large Earth microbes such as algae and protozoa; this we suggest, is *a priori* evidence that our BEs are incoming from space to Earth. One of the reasons we sampled over the grass-rich Prairies was to increase the likelihood that we would sample pollen and grass shards, but yet again, we found nothing but the unusual BEs on our sampling discs.

The response to the publication of our first paper on BEs was predictable Occam's razor was uncritically evoked to explain our findings. William of Occam's famous dictum states that, given two possibilities, the obvious is likely to occur. So, if we see and film an elephant in our local supermarket carpark we might suggest that it either came from space or escaped from a passing circus. Occam's razor tells us that the likelihood is that the elephant originated from a circus. At first sight, this seems to clinch the matter, but Occam's dictum can over simplify matters leading to a wrong conclusion (as the saying goes "you can cut yourself on Occam's Razor"). What if, for example, we investigated further and found that no circuses passed with a thousand mile of where the elephant was seen? This fact would not of course prove that the elephant has a cosmic origin, but it would show that you cannot use Occam's razor naively without reference to the complexity of a situation. Francis Crick of DNA structure fame, has firmly stated that Occam's dictum is not applicable to biology, yet it was blindly applied as a criticism of our work without any reference to other evidence which suggest that our BEs have a space origin.

Our work soon covered the internet and we were informed by the administrator responsible for public relations that it was the biggest media splash that the University of Sheffield ever had. The person in question was a young girl who was new to the job, and thinking she was doing her job and promoting the University. Unfortunately, "the suites" at Sheffield apparently reprimanded her; certain members of the University obviously thought that claims that life originate from space brought the University into disrepute! The administrator left to take maternity leave and we never found out what happened to her. We hope that the apparatchiks at Sheffield were kind to her.

This would not be the first time that senior members of the University of Sheffield would set out to actively block MW's work, the result being that MW went through a trying period of critical attack against his academic freedom.

An excellent example of how much prejudice there is against the idea that life comes from space is provided by what happened to former PhD student Tariq, who is one of the best research students MW has had over his forty-year career. Some years ago, Tariq gave an excellent talk on his research on panspermia to a UN space meeting. Afterwards he was accosted in the corridors by a very senior member of the European Space Agency who advised him that if he wanted a serious job in science he should immediately ditch MW as his supervisor and switch to a different research topic. When Tariq told MW this story his initial response was one of fury, who the hell did she think she was! On calming down however, he realised that she was giving "tough love". She herself did not need to be a bigot to pass on the truth about what scientist-bigots can do to student's future. Tariq returned to his native Iraq and made a fertile career in administration. If this episode truly reflects modern science, and we are afraid it does, then Tariq is well out of it! We rant on about the injustices which the two present authors and anyone else working on panspermia have had to suffer. They are real, but in the scheme of things they are minor insults, and they can be casually dismissed as victimhood, so let us get back to what matters — the sheer joy of scientific discovery, and of course the pursuit of truth.

The Sheffield-based Stratosphere Balloon Sampling

Material was obtained from the stratosphere using a specially constructed sampler, lofted by a helium weather balloon, between 2023–15. The first balloon was launched in July 2013 from Chester, Cheshire, followed by other UK launches, one in Iceland and one over Death Valley and finally, Wyoming, USA. The balloons reached heights around 22–28 km and the sampling times, when the sampling-drawer was open were of the order of 25–145 minutes. The sampling box, which was lifted by the balloon, contained the sampling apparatus (Fig. 13.2) which is described above; much more sophisticated versions of this sampler, improved by 3D printing are now being used. Sterile carbon discs were included in wells in the drawer. These adhesive carbon tabs, sometimes referred to as Leit tabs could be removed after landing and placed directly into a scanning electron microscope. In this way, we could collect any particles as the sampler crossed the stratosphere, over the time span when the drawer remained open. The sampling apparatus was shielded from the possible downfall of particulate matter from the balloon itself by means of a cover.

Before the launch, the sampling drawer was scrupulously cleaned, air-blasted, and swabbed with seventy percent alcohol to ensure that no remaining contaminants were present. A video camera was also included in the sampler box in order to monitor the opening and closing of the sampling drawer in addition to recording the view of the Earth from the stratosphere. An instrument which records various data such as GPS position and altitude, internal and external temperature, humidity, air pressure, acceleration (multi-directional), and magnetometry was also included. After release, it ascended to upper tropopause boundary, entering into the stratosphere, where the low pressure eventually caused thinning and expanding of the balloon's elastic material, eventually leading to it to burst. A parachute was then deployed to slowly bring the box safely back to Earth. On landing, the box was examined to check for any damage. Even the slightest damage meant that no further analysis of the samples took place. In addition to particulate samples, we also obtained some exquisite images of the stratosphere (Fig. 13.3).

Control Flights

Separate Control Flights were also performed before each of the sampling flights, where the drawer box remained unopened, but all the other analysis techniques were followed in the same manner as the test launches. No particulate matter from Earth was ever found inside the sampler, proving that the drawer was sealed airtight. As

Fig. 13.3. View of Death Valley taken from the ascending sampler.

a result, none of the stubs were exposed to any terrestrial particles which might reside at ground level or any height in the stratosphere. The negative findings from the controls also prove the sterility of the samples' processing and analysis procedures.

What We Found in the Stratosphere

When the balloon, with the sampler, ascended above 22 km, the sampler drawer shown in Fig. 13.2 was opened outwards, thus enabling the SEM stubs mounted on it to be exposed to the conditions of the surrounding stratosphere, and allowing particles that might be present in the air to fall on the mounted stubs We also located stubs on the outside of the sampling box to show the difference between stubs exposed to transit up and down through the atmosphere to those left in the sampler. As we expected, we found lots of Earth-derived debris on the outside of the box, including pollen, spores and grass shards (Fig. 13.4). An absolutely crucial point, worth re-iterating, is that such particles were never observed on any of the SEM stubs positioned inside the drawer.

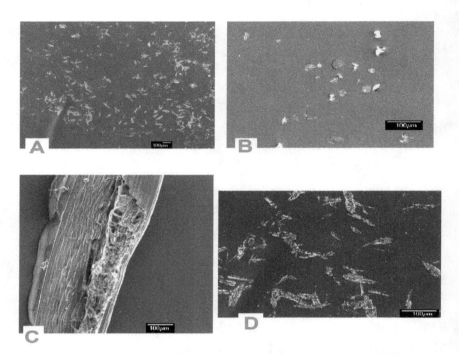

Fig. 13.4. SEM images of terrestrial material found on the outside of the sampler box. A. Pollen and fungal spores and B, C, and D, part of a grass shard found only on stubs placed on the **outside** of the sampler.

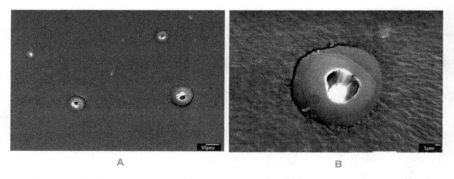

Fig. 13.5. Marked impact on the sampling discs resulting from high-speed, incoming micrometeorites are shown in A and B.

When we examined the first disc which was exposed to the stratosphere we were delighted to see that craters appeared on the hard carbon discs, presumably caused by cosmic dust impacting the sampler (Fig. 13.5).

Figure 13.6. shows how essential it is to employ EDAX elemental analysis to back up any claims that a microscopic particle is a true BE. This object could easily have been interpreted as a "mouse-like", biological entity, with legs front and rear and a "snout" on the front (right) side. The elemental analysis below however, shows that it contains none of the carbon and oxygen (C and O) typical of life forms and the other BEs, but a single, marked peak for aluminium (Al). This is in fact, a stratosphere-derived "mouse" made up entirely of aluminium; it is a mouse that never lived anywhere!

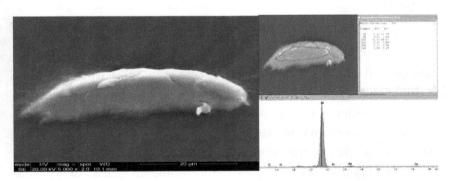

Fig. 13.6. A particle that looks biological but EDAX shows is inorganic and made up of aluminium.

It is crucial to note therefore that all the BEs described below are comprised of carbon and oxygen, with the occasional EDAX signature for nitrogen and do not contain large amounts of mineral elements, such as calcium, silicon, and iron which are typically found in cosmic dust, incoming to the stratosphere (the exception being the diatom frustule made up entirely of silicon, and the stratosphere-derived sphere which contains titanium and a slight amount of vanadium).

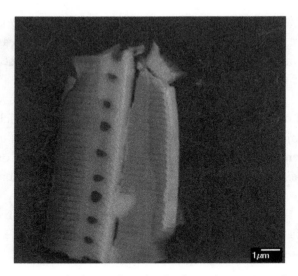

Fig. 13.7. Fragment of a diatom frustule, closely similar to a *Nitzschia* species.

We were delighted to see an obvious biological entity, collected from around 24–28 km in the stratosphere (Fig. 13.7). There could be no doubt that this is a broken diatom frustule, resembling *Nitzschia* a terrestrial species which lives in freshwater and the oceans. Diatoms are ancient organisms; whose evolution we know little about. They consist of an often beautifully sculptured silicon shell which contains the living organism (or protoplast). The protoplast gains energy from sunlight, while fixing carbon dioxide by photosynthesis. What we see here is the broken silicon shell, the internal protoplast having long since disappeared.

When news of our first paper on BEs hit the media a number of scientists immediately suggested that this diatom frustule must have come from Earth, and that our claims that it was incoming from space were nonsense. Most critics simply invoked Occam's razor and said that the most obvious explanation was that we had caught a diatom as it was uplifted from Earth. These naysayers clearly had not read our paper, because had they done so they would have found that we discussed and dismissed the Occam Razor argument.

We argue then that the diatom frustule came from space, probably from a watery planet or comet. Watery planets are common in the Universe and these could provide a home, as could comets, for oxygen-producing, photosynthetic diatoms. Diatom frustules would act as an ideal physically protective panspermic-vehicle for the algal cell itself, and for any oxygen requiring associated organisms, such as an aerobic bacterium; frustules might also protect against ionizing radiation. In addition, diatoms are known to be resistant to UV and B, although possibly not UVC; protection against UVC would however, be provided by the finest smear of organic material or inorganic salt originating from the home, watery environment. The input of a large quantity of oxygen-producing photosynthetic organisms would obviously be highly advantageous to an anoxic, early Earth. It is highly noteworthy, that as mentioned above, we have observed diatom frustules in a meteorite, adding weight to our claim that they exist elsewhere in the cosmos. Encasement in these bolides would provide additional protection for diatoms, and other microbes, as they journey through the cosmos. Richard Hoover has pointed out that there exists a close relationship between the measured infrared properties of diatoms and the infrared properties of interstellar dust present in parts of the cosmos, and that many diatoms are the right size and shape for entry into the Earth's atmosphere, even at diameters up to 100 microns; finding which again point to the panspermic origin of the Earth's diatoms.

Amazingly, we continued to isolate more and more unusual forms from the stratosphere, with every launch, but one, yielding BEs. Even this occurrence fits expectation, since we would expect these, what we believe are organisms, to exist in the stratosphere in "swarms", so that on some sampling trips they will be missed and nothing will be sampled.

Let us look at some of the more interesting BEs. Figure 13.8 shows a somewhat phallic image. One of the consequences of examining living things under the scanning electron microscope is that spheres and tubes tend to collapse under the vacuum which has

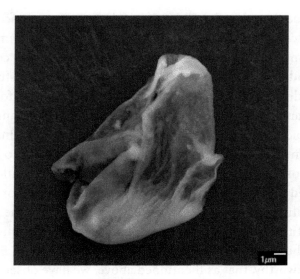

Fig. 13.8. A somewhat phallic BE isolated from the stratosphere.

to be imposed, the structure seen here is therefore in reality balloon-shaped. The UK, popular newspaper referred to this particle as being "an Alien's manhood!"

In the middle of the balloon structure, we see a proboscis with a nose-like opening, which would clearly have extended downwards. There is an opening or sphincter at the top and an ear-like projection on the side. We assume that this organism lived in a watery environment (possibly on a comet). The proboscis, we speculate, would have been used for feeding and the sphincter for ejecting waste; the ear-like structure we think will be replicated on the unseen side of the balloon and is possibly used for propulsion.

Without doubt the most significant BEs we have isolated is shown in Fig. 13.9. It is an almost perfect sphere which the EDAX report showed is made up mainly of carbon and oxygen, but also titanium.

Whether or not this element is present as titanium metal, or a compound, like titanium oxide is not clear. A close look at the sphere showed it was covered in what looked like mycelium. So, a perfect sphere showing signs of biology had been captured from the

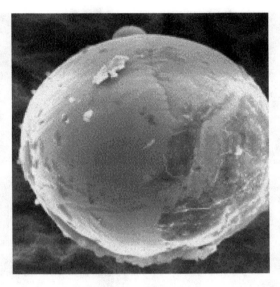

Fig. 13.9. A sphere isolated from the stratosphere made up mainly of carbon and oxygen, but also containing titanium. Note the surface filaments to the right.

stratosphere at around 24–28 km. More excitement followed when Chris and Alex were asked to go back and try and move the sphere over. MW is not certain to this day why he made this request, but when they used incredibly sensitive needles to move the sphere under the scanning electron microscope they revealed an impact crater, and material oozing from out of the sphere (Fig. 13.10). The fine needles were seen to be bent, showing that the sphere had taken some moving from its original position.

This, together with the presence of an impact crater in the hard carbon disc shows that the sphere was travelling at speed when it collided with sampling disk. It could not have travelled at such speed coming up from Earth, so the presence of the impact crater provides strong evidence that the sphere was incoming from space when it hit the sampler. Analysis of the material oozing out of the sphere (Fig. 13.11) showed it to be made up only of carbon and oxygen, that is, the ooze is biological.

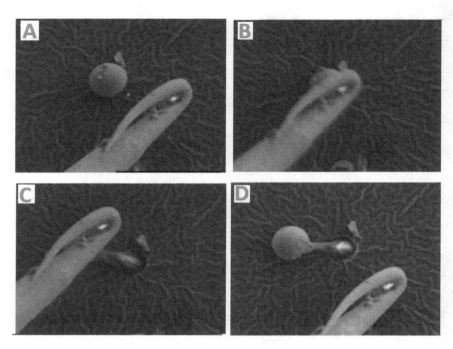

Fig. 13.10. SEM images for the Large Sphere. The sphere was moved using nickel-chromium nano-tips. A: Shows the LSO during initial attempts to remove it, B, and C: during the manipulation, and D: after removing the needle away; note the biological material oozing out.

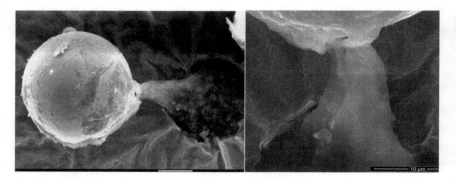

Fig. 13.11. The titanium-rich BE after being moved across sampling stub. Showing a filamentous outer coat and carbon-oxygen-rich material oozing out next to an impact crater.

To summarize then, we isolated from the stratosphere a relatively large (in microscopic terms) an almost perfect sphere, made up of carbon, nitrogen and titanium and covered in what looks like biological mycelium; the inside of the sphere contains a biological medium made up only of carbon and oxygen and traces of nitrogen. We are bold enough to assert that the isolation of this sphere and the other BEs shown above proves that life exists elsewhere in the cosmos and is continually incoming to Earth. We are not the only life forms in the cosmos!

What will critics make of the extremely bold claim? Well firstly, we know that similar spheres are found on Earth, so the obvious claim will be that our sphere is terrestrial. However, as with the other BEs we have isolated, we cannot find descriptions of a similar

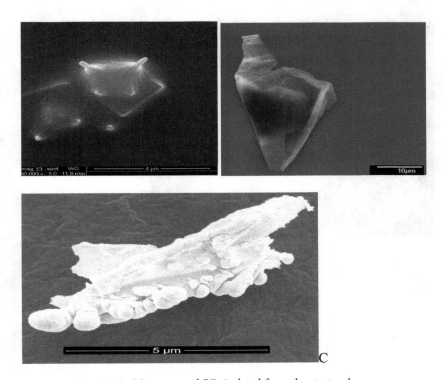

Fig. 13.12. More unusual BEs isolated from the stratosphere.

biological sphere being found in an Earth environment. Geology-derived spheres always contain large amounts of elements like calcium silicon or iron, while man-made spheres (like coal ash and other fuel-burning derived spheres) are rich in carbon, oxygen, but also contain large amounts of mineral elements, such as calcium silicon, and iron, which are not seen in large amounts in the stratosphere-derived sphere. It is difficult to suggest a mechanism which could lift such a relatively large microscopic sphere from Earth to a height of between 24–28 km. No such mechanism has yet been identified.

More unusual BEs are shown in Fig. 13.12. The "horned" BE is smaller than five microns, but is attached to a salt crystal which could not have been elevated to the stratosphere. The flask-shaped organism

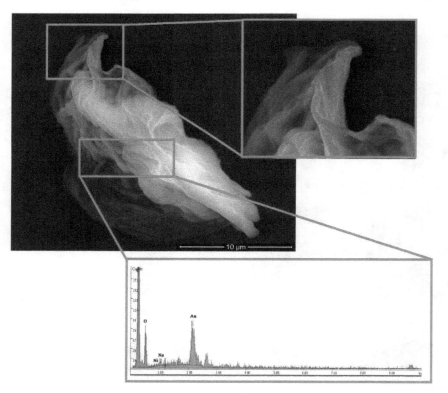

Fig. 13.13. A gossamer-like particle recovered from the third stratospheric sampling trip.

is partially collapsed due to scanning electron microscope treatment, and the highly unusual BE shown in the bottom image is either part of single organisms, or is a complex of smaller organisms attached to a "base plate".

An unusual gossamer-like organism is shown in Fig. 13.13. Note the thick ridge in the inset image, which suggests that the gossamer is not a paper-thin sheet, but a collapsed balloon-like structure. The EDAX inset at the bottom shows a prevalence of C and O.

Very surprisingly, we also found, what look like biological, filaments attached to a mass of inorganic particles; in fact, the filaments help to bind the mass together (Fig. 13.14).

Fig. 13.14. Detail of a filament-rich BE isolated from the stratosphere. EDAX of filaments show them to be made up only of carbon and oxygen.

We are surprised because such filaments could not survive, an unprotected journey from Earth to the stratosphere. The filaments are clearly attached to the particle mass and are not just air borne laboratory contaminants, like microscopic cotton threads. We also know that the filaments are not mineral hairs because they EDAX at carbon and nitrogen without silicon and heavy metals. Areas where the filaments collapse were also seen; fungi and other filamentous organisms show this response when exposed to the vacuum imposed during scanning electron microscopy. Ergo these are biological filaments. Since such filaments will be susceptible to breaking when traveling through space, so we assume that the particle mass shown here would have been part of a larger mass which would have afforded protection to the filaments.

Finally, Fig. 13.15 shows an object impacting the graphite sampling disc at its base. The particle is around 50 × 20 µm in size. It is clearly not a piece of inorganic dust but appears to be biological in nature. It is not obvious however, if this object is a single entity or part of a larger biological structure or, indeed, a group of smaller forms massed together. It possesses distinct spiracle-like openings. The damage at the base, we assert is caused to the sample disc surface the BE-carrying ice particle impacting the sampler.

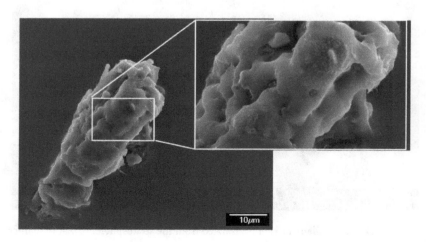

Fig. 13.15. A BE with spiracle-like pores.

How Can we be Certain that Biological Entities Originate from Space?

Before we answer this question we need to gather together the evidence that the stratosphere-derived biological entities are in fact organisms. Well, firstly they look like organisms! Without additional evidence a critic would scoff at this suggestion, simply because, as we have mentioned before, our minds have a passion for seeing pattern in random images (so called pareidolia). Every now and then newspapers publish pictures showing seemingly unbelievable images, such as the face of Christ seemingly appearing in the draining's of a beer glass! Evolutionists think humans developed this ability in order to immediately recognise patterns so that we could avoid predators. Even if a random image is not a lion we had better see it as such and react accordingly!

The BEs range from around 10–40 μm plus and clearly exceed the "5-micron rule" and could not (on the basis of present modelling) have been elevated from Earth to space. EDAX analysis shows that BEs are composed of carbon with trace of nitrogen and lack elements such as silicon, calcium and heavy metals typically found in cosmic dust. These stratosphere-derived organisms are highly unusual and cannot be assigned to terrestrial taxonomic groups. BEs exhibit bilateral, symmetry, show varied morphology and do not, we maintain, false sightings resulting from pareidolia.

For safety reasons, we did not attempt to culture any of the stratosphere-derived BEs. We are frequently asked if these organisms are dangerous to life on Earth. Since they have been probably incoming from space to Earth for aeons they are as much part of biology of our planet as we are. It is possible however that some of the stratosphere-derived organisms could pose a risk; as indeed could incoming bacteria and viruses. The possibility that pathogens, including the SARS and the corona virus could have a space origin is discussed in the theory of pathospermia and has been emphasized by the work of Hoyle and Wickramasinghe and more recently by Wickramasinghe and co-workers.

If microscopic life forms have been incoming to Earth from space for aeons, as suggested by our findings then it is possible that their DNA may have been infected with life from other planets and played a major role in the evolution of life. This possibility reached its apotheosis with the suggestion by Professor Ted Steel of Chandra's group that octopus genes and the Cambrian explosion arrived from space; the claim appeared in a suitably peer–reviewed journal, but not surprisingly, has not been readily accepted.

The fact that life forms impact Earth means that they will also impact other astronomical bodies, including our Moon. Why then do we not find extensive deposits of dead BEs on the moon's surface? The answer is simple, the Moon has no atmosphere, so any incoming BEs would be smashed to smithereens and any carbon associated with them would be destroyed by UV light incident on the moon's surface. According to our theory, the origin, development and evolution of life on Earth could only have occurred following the development of an atmosphere sufficient to slow down incoming BEs and other life forms. Life will therefore be spread around the cosmos, but will only be established on planets which have similar atmospheres.

A very large number of cosmic dust samples have been captured in the stratosphere by NASA and others. Is there any evidence of biology in such samples? We can only find two samples from the images posted on the Internet, which we think are relevant to our studies; these are shown in Figs. 13.16 and 13.17.

The first cosmic dust image, published by NASA, shows a mass of cosmic dust particles and filaments — which could be biological, at the top left and bottom right. An amazingly similar image of filaments present on a particle mass isolated from the stratosphere by us, is also shown (Fig. 13.16). A far more interesting image, this time of an interplanetary dust particle, is shown in Fig. 13.17; published by Luigi, Colangeli, components of the particle mass look distinctly biological.

This again shows a mass of particles and large collapsed balloon-like structures. There are numerous filaments apparently holding the particles together. In the bottom right of the image it can be seen, that one of these filaments overlays and cuts into one of the balloon-like structures. The collapse of the larger balloon-structures is the result (as we have frequently seen), as we keep mentioning, of imposing a vacuum during the use of the electron microscope. These samples could have been badly preserved in a moist environment, allowing terrestrial biology to grow on them. This seems very unlikely however, since cosmic dust curators go to great lengths to store their samples in the appropriate manner. Should the veracity of these images be questioned it would not of course have a negative impact on our findings.

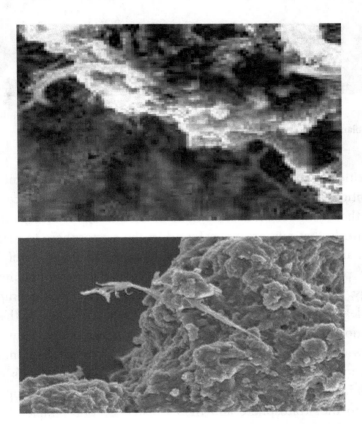

Fig. 13.16. A NASA cosmic dust particle, showing biology-like filaments (top image) and a remarkably similar filament-rich particle mass isolated by us from the stratosphere (bottom image).

Fig. 13.17. An interplanetary dust particle showing signs of biology (with permission of Luigi Colangeli).

Critics often quote Occam's Razor, or parsimony, to dismiss our contention that our BEs are incoming to Earth from space. They argue that since the Earth is home to a vast amount of biology it is far more likely that any organism found in the stratosphere to originate from Earth, rather than space. While Occam's Razor is a useful tool in the physical sciences it can be a very dangerous implement in biology. As Francis Crick pointed out, it is very rash to use simplicity and elegance as a guide to biological research. One of the first indications that biology resides in the stratosphere came, as we have seen, with the isolation of a diatom frustule from this region. Since vast numbers of diatoms exist in the world's oceans critics could, and did, argue that this frustule must have been carried up from Earth to the stratosphere and then isolated by our sampler In fact, the isolation of an empty diatom frustule in the stratosphere adds, rather than detracts from our theory. Watery planets are common in the Universe and these could provide a home, as could comets, for oxygen-producing, photosynthetic diatoms. Diatom frustules could act as an ideal physically protective panspermic-vehicle for the algal cell itself, and for any oxygen requiring associated organisms, such as an aerobic bacterium; frustules might also protect against ionizing radiation. In addition, diatoms are known to be resistant to UV and B, although possibly not UVC; protection against UVC would however, be provided by the finest smear of organic material or inorganic salt originating from the home, watery environment. The input of a large quantity of oxygen-producing photosynthetic organisms would obviously be highly advantageous to an anoxic, early Earth. It is highly noteworthy that claims have been made for the isolation of diatom frustules from presumptive carbonaceous meteorites which would provide additional protection for these organisms as they journey through the cosmos. Finally, Hoover reported that there exists a close relationship between the measured infrared properties of diatoms and the infrared properties of interstellar dust present in parts of the cosmos, and that many diatoms are the right size and shape for entry into the Earth's atmosphere, even at diameters up to 100 μm; results which again point to the possible panspermic origin of the Earth's diatoms.

Should any stratosphere pollen grains be isolated, either by us or other workers, then this would be taken as evidence that all of our sampled BEs must originate from Earth. However, according to our current understanding, particles greater than five microns in size cannot cross through the tropopause into the stratosphere. This claim relies on the work of James M. Rosen, who found a "natural break" in the distribution of stratospheric particles; anything five microns, or below, is of terrestrial origin, while anything above this is cosmic dust and incoming from space. C.E. Junge also concluded that terrestrial particles larger than one micron are only found in the stratosphere for very short periods following rare, major volcanic events, like Krakatoa. If a pollen grain or other Earth-derived particle, smaller than five microns reached the stratosphere then this would not contradict our suggestion that BEs originate from space. If, however, a pollen grain larger than five microns were to be found in the stratosphere then this would demonstrate that an, as yet unknown, mechanism exists which can elevate particles bigger than this across the tropopause. Such a finding would at first sight destroy our claim that the BEs which we find in the stratosphere originates from space. In this case we would initially fall back on claims of organism novelty, i.e. that since our BEs (with the exception of the diatom fragment) are morphologically similar to no known biology found on Earth they must originate from elsewhere, an obvious tenuous position since such stratosphere-derived biological entities may exist, undiscovered on Earth (one would still however, ask why it is that common, known organisms, are not over represented on our sampling stubs). The claim that we have discovered biology originating from space would however, continue to be supported by the fact that, in some cases, our stratosphere-isolated biological entities cause impact events on the sampling stubs (see for example, Fig. 9), and in the case of a biology-associated titanium ball, a clear impact crater (Fig. 8). We assume that such impacts are caused when the BEs reach the surface of the sampling stubs from space inside minute ice grains; it is these we theorize which cause the impact events, only to disappear and leave the pristine BE resting on the impacted carbon surface of the stub. Such impact damage could not be caused by organisms "lazily" drifting up from Earth. They might however, be explained on the basis that an unknown elevation mechanism can carry five micron-plus BEs and particle masses, such as the ones we have sampled, from Earth to extreme heights in space.

We remain confident that the evidence continues to show that the increasing number of biological entities which we have isolated from the stratosphere, some of which are described here, originate from space, most likely from comets.

14

Conclusions

A s we have seen, the basic concept of panspermia has a long history stretching back over many centuries or even millennia, but until the1960s, when NASA made some tentative exploration, no attempt had been made to provide any experimental evidence in support of the theory. The reasons are obvious. Prior to this date it was difficult to reach and to sample the stratosphere and then investigate, using electron microscopes and other techniques (like EDAX), anything that was captured there. Even when these difficulties were overcome the stratosphere was assumed to be sterile and little interest was shown in exploring the possibility that this region is home to a distinct biosphere. The studies of the Russian, A.A. Imshenetsky should have alerted scientists to the reality that microbes can be found at heights above 42 km, but his work was, as usual, dismissed on the basis of contamination. Work by the Indian group and ourselves later confirmed the Russian studies, as we discussed in earlier Chapters, but the same old excuse was used in order not to take these findings on-board. There have now been sufficient independent studies to show that microorganisms do indeed exist in the stratosphere. The question then is how did they get there? Surprisingly, even experienced scientists claim that "they just drift up there". In reality, it is extremely difficult to elevate even small, one micron, particles to the high stratosphere. We of course assert, that it is impossible for the larger (10 to 40 μm plus) particles, which we find in this region, to come from Earth. Instead, our evidence shows that they are incoming to Earth from space. Such findings show that the theory of panspermia is the only theory that can accord with the facts that life exists elsewhere in the cosmos. Of course critics might be emboldened to say, without any proof, that large particles can be elevated to the stratosphere, but the mechanism by which this is achieved has yet to be discovered. Obviously, we cannot argue against such strident soothsaying. However, the evidence that clinches it for us, is the absence of common terrestrial organisms on the inside of our samplers, such as, fungal spores and grass shards; all of which we find on the outer sampler containers which are exposed to the lower stratosphere. Since there is no sieve between this region and the stratosphere capable of holding back these common organisms, and only allow our BEs to go higher, then these unusual organisms we assert are arriving at the Earth from space.

If microscopic life forms have been incoming to Earth from space for aeons, as suggested by our findings, then it is possible that their DNA may have played a major role in the evolution of life. This possibility reached its apotheosis with the suggestion by Edward Steele, CW and collaborators that octopus genes and the Cambrian explosion itself arrived from space; the claim appeared in suitably peer reviewed journal, but not surprisingly, has not been readily accepted; could all the defining genes of novel life forms have indeed arrived directly from space?

Another question often raised is why do we not find extensive deposits of dead BEs on the Moon's surface? The answer is simple, the Moon has no atmosphere, so any incoming BES would be smashed to smithereens and any organic molecules associated with them would be destroyed by UV light incident on the Moon's surface. According to our theory, the origin, development and evolution of life on Earth could only have occurred following the development of an atmosphere sufficient to slow down incoming BE's and other life forms. Life will therefore be spread around the cosmos, but will only become established on planets which have similar atmospheres to our own Earth. Of course, there are always "flies in the ointment" of all scientific findings which need to be addressed. We have two. The first of course is the fact that we isolated an empty diatom frustule from heights of 24–28 km. Diatoms are common Earth organisms, occurring in untold numbers in the Earth's oceans. Our stratosphere-isolated diatom could therefore have been simply transferred from this environment to the stratosphere. The problem with this argument is that a diatom frustule is even less likely to make this journey than are, for example, lighter, grass particles.

We realise that here we are risking becoming "hostages to fortune" since if any stratosphere pollen grains, or fungal spores are isolated, either by us or other workers, then this would be taken as evidence that all of our sampled BEs must originate from Earth. (The Frontispiece image, top line, right shows a particle which could possibly be a pollen grain of the garden flower, hollyhock, *Alcea* species.) This realization leads us into the second, and most damaging fly in our ointment that is provided by the work of Della Corte *et al.* in 2014. They claim to have isolated a complex, unknown biology-like particle at circa 24–28 km in the stratosphere, which they described as being non-mineral and comprising mainly carbon and nitrogen, i.e. by our definition a Biological Entity. They also reported finding a chain of fungal spores at this height. At first sight this latter claim is, of course, fatally damaging to our argument that terrestrial spores, etc. do not, because of their size, reach the stratosphere. However close examination of the spore chain provided by Della Corte shows that it is unbroken and pristine and therefore is unlikely to have survived as a chain in the stratosphere, an environment of harsh winds and cosmic dust movements. In fact, fungal spore chains are never found when fungal-spore collectors are used to sample the relatively calm Earth's atmosphere; only isolated spores of *Cladosporium* species are found, never pristine chains. It is noteworthy then that Della Corte suggests that the relatively benign effect of the SEM beam detached a fungal spore from the parent chain. The suggestion that such a spore-chain is of terrestrial origin and was isolated in the stratosphere is therefore untenable. As a result, the above-mentioned findings do not fundamentally damage our claim that the BEs we isolate in the stratosphere originate from space. Ironically, after suffering from suggestions that all of our work results from contamination, we use this argument to protect ourselves. Sensibly argued claims rather than knee-jerk reactions about contamination are however, acceptable.

We know that volcanic dust, ejected by rare, super eruptions, can reach the lower stratosphere. Could this dust be contaminated with Earth biology? Since ejected volcanic dust is super-heated, we think this is unlikely. It could be argued however, that such dust picks up BEs as they ascend through the lower atmosphere. Why, we ask, would these organisms be unusual, and how could organisms around forty microns, and larger, in size be picked up, and elevated by sub-micron volcanic dust particles?

The reader will perhaps have noticed that we have not mentioned much about viruses. This neglect is simply because we have no means of detecting viruses in the stratosphere. It would also be difficult to demonstrate that viruses are incoming to Earth from space, simply because being so small, they will be continually uplifted from Earth. We have however, stated that viral diseases may originate from space and that incoming viruses may add information to the Earth's biome which will have an influence on evolution.

As we have mentioned above, a number of critics have stated that if our work is correct, then the surface of the Moon and other planets should be covered with a layer of diatom fragments and other biological entities originating from space; clearly this is not the case, not at any rate for the Moon. We have provided an obvious answer to this apparent conundrum, namely that the Moon does not support an atmosphere, and as a result, any biological particles impacting their surface would be obliterated. The Earth, in fact, provides a "Goldilocks' atmosphere" which is "just right" for demonstrating neopanspermia.

We are also plagued by the application of the aphorism "extraordinary claims require extraordinary evidence". Implicit in this statement is the idea that a novel scientific idea can only be suggested when **all** the information demonstrating it to be correct is available. Science does not however, progress in this manner, but instead proceeds by the accumulation of small increments of new knowledge. We prefer another of Carl Sagan's quotes namely that "somewhere, something incredible is waiting to be known."

Where do we think our biological entities originate, and how do we think that they got here? The Hoyle and Wickramasinghe theory of cometary panspermia lays great stress on comets, hence it is being referred to as the "Cometary Theory of Panspermia". Comets provide ideal conditions for the growth and replication of microbes, and the cometary theory suggests that these bodies act as ideal incubators for microscopic life forms. These life forms can then be spread throughout the cosmos by well attested astronomical processes in the manner that has been extensively discussed by CW, Fred Hoyle and their many collaborators and co-workers over many decades.

It is worth noting that the stratospheric balloon work of one of us MW has revealed the presence of craters on sampling discs that were exposed to the stratosphere caused by impaction by cosmic dust particles. Occasionally, we also see such impacts around BE's (putative biological entities). Now these organisms are much less dense than cosmic dust, so how do they form impact craters? We believe that the comets, containing BE's, break up into smaller clumps which by the time they reach the samplers are minute particles; it is these particles which impact the sampler, causing the formation of a crater. When the cometary ice particles evaporate, then any BE contained within is left intact and comes to be associated with an impact crater.

Another question we are often asked is — Do you have any evidence for Directed Panspermia? This idea relates to the suggestion that life was intentionally sent to Earth by some unknown cosmic civilisation. The idea would likely have never got traction had it not been championed by the Nobel Prize winning biologist Francis Crick. Presumably, life would have been sent (and possibly continues to be delivered) by some kind of craft. It has even been suggested that it might have been derived from the waste products of visiting aliens! Since we are unlikely to find such a delivery vehicle the idea remains just that, an idea. For the sake of completeness, MW has suggested that BEs, notably the titanium spheres could act as a vehicle for directed panspermia. Not surprisingly the internet is full of conspiracy theories that these spheres could be spread by Governments of Earth, in order to spread some possibly toxic agent. Strangely, of all the countries in the world such conspiracy theories seem always to emphasize the USA, and without any evidence the involvement of the CIA or NASA. A single rocket containing a payload of microorganisms would effectively inoculate a single, lifeless planet, but perhaps a better idea would be for the alien experimenters to inoculate comets with microbes and let them randomly spread life wherever they travel.

Of course such speculation begs the question why would aliens bother to do all this? Well, we bother to do all kinds of experiments, just for intellectual gratification; aliens might — just might — be far more intellectually developed than us, so maybe directed panspermia could be simply the work of a few alien hobbyists, playing with life in the universe!

It is noteworthy that we also demonstrated the presence of DNA in some of the biological entities. Recent studies discussed in a peer reviewed PhD Thesis by Tariq Omairi show that a vast array of bacterial DNA is present at around 24–28 km in the stratosphere. Tariq additionally isolated human DNA from stratosphere samples. Although the obvious possibility is that such DNA is a contaminant, this finding (should it be confirmed) opens up the intriguing possibility that small particles of human material such as sub-5 micron skin-fragment are being continually elevated from Earth to the stratosphere, and possibly beyond.

The fossilized biology we find in the Polonnaruwa stone is without question truly amazing. We provide convincing evidence that this stone is a meteorite, but it is yet to be officially accepted as such by the US-based Meteoritic Society, a fact which will be used to summarily discount our work. The important point however, is not whether the stone meets the accepted definition of a meteorite, but that it is *not of terrestrial origin*. Biology may be extinct in the home of the Polonnaruwa stone, but this need not be the case; after all, we live on a planet teaming with both living things and dead fossils! A different issue arises in relation to the North West Africa meteorite. Here, we know we are working with an officially recognized meteorite, but the images of biology it contains, although convincing, are less dramatic than those found in Polonnaruwa sample.

Finally, it is important to note that any individual part of our findings could be proven wrong without compromising the theory of panspermia. For example, BEs may be found to be terrestrial organisms and the meteorite fossils that we, and others, find may turn out to be artefacts. Such individual findings would not fatally compromise our varied evidence that life originated from space and continues to arrive from this source.

Our conclusions can be summarized as follows:

(1) Bacteria and microscopic fungi are present in the stratosphere at heights of at least 41 km. Individual organisms of size around one micron and below, could possibly be elevated to the stratosphere, but not clumps of bacteria around 10 μm. We suggest, that the stratospheric bacterial flora consists of two populations, one mostly exiting and a second, incoming population.

(2) Large DNA-containing masses are incoming to Earth from space.

(3) Complex biological entities are incoming to Earth from space. We suggest that these originate from comets. The Entities are presumably mostly dead, although a small proportion could be living. Our claim that these relatively large organisms are incoming from space is based on (a) the "five micron rule", that is no biology above this size reaches the stratosphere from Earth, (b) the absence of common Earth-derived material inside our samplers

and (c) the fact that some particles are associated with impact events on the sampler discs, caused we think, by cometary ice-impact. An alternative explanation is that the BEs, we find, were originally Earth-based organisms, which were ejected from our planet by a massive impact event (such as the one believed to have made the dinosaurs extinct) and are now returning.

(4) We suggest that all incoming biological material could contribute information to the Earth's biome and therefore influence terrestrial evolution.

(5) We have not studied viruses, but it is likely that these are also incoming from space to Earth.

(6) These incoming organisms are as much part of Earth's biology and we assume most will be harmless. Some however, (notably viruses) could cause pandemics.

(7) Meteorites contain fossilized organism showing that microbial life once existed on Mars and other planets.

Finally, let us return to one of the Biological Entitles discussed above which is particularly significant (Fig. 14.1). The final figure is a repeat showing of a large particulate mass made up of inorganic cosmic dust particle interwoven with filaments which, because they are not inorganic, we suggest are biological. When MW first saw this image he was a little concerned thinking that it was unlikely that filamentous organisms would make the journey from space to the stratosphere. But this particulate mass provides us with a smoking gun that life is incoming from space. Firstly, look at the size. There is no way that a mass of 2–300 μm could be elevated to just below 30 km.

Secondly, if the mass originated from either space or Earth without being protected, it would have been broken up and the filaments would have been destroyed. This points to the fact that the mass was in some way covered and protected. If it originated from Earth with such a covering, it would be far too heavy to be elevated to the stratosphere. If it came from space however, a covering would

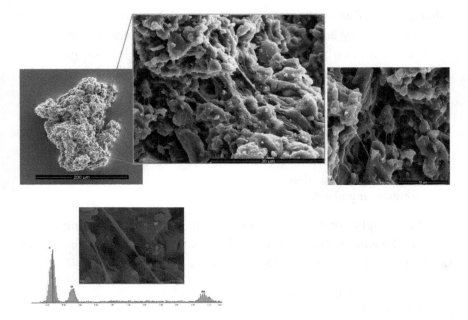

Fig. 14.1. A filament-rich BE 200plus microns in size. Note the EDAX showing that the filaments are biological.

not present a problem, and in fact would protect the particulate mass. We believe that this covering is cometary ice, which evaporated leaving the filamentous mass in a pristine condition. This particulate mass contains biological filaments and came from space. Yet again we provide evidence that Earth is not the only place in the cosmos that supports life.

Our conclusions will doubtless be met with much hand-waving and the "internet-Einsteins" and trolls will very much enjoy attacking them. More rational scientific debate may also dismiss our claims. We ask however, that rather than endlessly debating our findings, the scientist should send samplers to the stratosphere to look for our BEs, or search meteorites for fossilized life forms. Neither of these tasks is expensive or time consuming, certainly not when compared to the vast amounts of money NASA spends on searching for life on other planets. Rather than speculating on why we are wrong, a small amount of money and effort could confirm, or dismiss our findings.

We are modestly certain that our evidence shows that microbial life exists elsewhere in the cosmos; we are not alone. It would be tragic if our claims were to continue to be subjected to a knee-jerk dismissal, or merely become the subject of endless, fruitless debate. Even we find it difficult to believe what we have found, but in the immortal words of that famous detect fictional detective:

> "When you have eliminated the impossible, whatever remains, however, improbable is the truth."

We thought we might end on a light note by concluding in the style of Darwin, firstly when he added his famous quote at the end of the first edition of *On the Origin of Species*:

> There is even more grandeur in this view of life. A view, which sees life-forms not being original, or restricted to a nugatory globe, in the corner of an insignificant galaxy, but able to spread throughout the cosmos. Life-forms that grow and evolve on any sterile world they meet, or otherwise contribute to nature's development on already populated worlds.

And mirroring Darwin's "Tangled hedge quote" in later editions of *The Origin*:

> It is wonderful to reflect on the feeling of awe one gets when gazing at the milky way, and to know that it is teeming with a diversity of life, and the nucleotide-blueprints of future life. All to be continually spread throughout the cosmos, by a natural process, or one that is directed by some, as yet, unknown intelligence.

Darwin's famous classic revolutionised our ideas about how species originated. We hope that this book will revolutionise our view of how life originated on this planet and convince the reader that it continues to be delivered to Earth from the depths of the cosmos.

Postscript — Future Prospects

G enetic similarities of stratospheric bacteria as discovered by Wainwright and others to existing terrestrial genera have been cited as an argument to discount their possible space origin. However, if all terrestrial bacterial genera all have an ultimate space origin, homologies of the type found (e.g. the genera discovered by Shivaji and others) are to be expected and do indeed corroborate a space origin of all bacteria on Earth as discussed extensively by Hoyle and Wickramasinghe in numerous books and papers.

In order to take the matter further, and hopefully reach a decisive conclusion, further tests of the collected microbial samples would be desirable. One such test (as has been discussed in the book) involves the deployment of the rather rare laboratory resource — a Nanosims machine. This technique which has to be deployed directly on the original cells (before replication) will determine the isotopic composition of carbon, oxygen and other constituent elements within the individual bacterial cells, and if the composition turns out to be non-terrestrial, it is QED! Proof once and for all that panspermia — neopanspermia — is proved. The tight control of the relevant experimental resources worldwide has so far prevented us from gaining access to this equipment. The sceptic is thus left in a seemingly comfortable position to assert, if he or she so wished, that what we have found in the balloon samples in 2001 and 2009, as well as in the samples recovered more recently on the International Space Station, were terrestrial contaminants — being lofted to the stratosphere even against incredible odds. This is just one instance of control of science by a small, but highly influential, group of "gatekeepers" that is hindering current progress in Biology and which cannot be easily overcome.

A paradigm shift of the kind discussed throughout this book may thus be stalled for seemingly rational reasons based on either prudence or pragmatism:

(1) The proposed paradigm shift from Earth-centred life to cosmic-centred life might be seen as a threat to national security. The critic might assert that people may become so scared and emotionally unsettled that even the enforcement of law and order might pose a problem.

(2) To admit that we have supported a wrong paradigm may have economic repercussions in regard to the many large funding commitments that are already in place for exploring ideas based on a wrong premise.

(3) An inevitable demand that would follow for re-orientation of existing space programmes would have serious fiscal implications that would need to be kept in mind.

(4) To cope with ameliorating the worst effects of any future pandemics it would have to be deemed prudent — even necessary — to monitor the stratosphere for potentially lethal incoming pathogens (bacteria and viruses), and such a world-wide programme would require large budgets as well as new strategies in world health economics.

None of the above reasons would be good enough to halt the acceptance of a new paradigm which we believe is long overdue. From the earliest times primitive man appears to have had an innate perception of a connection between life and some major aspect of the Universe. The fact that most primitive gods and goddesses being invariably placed in the skies was itself an expression of this link. With the triumph of Christendom over Greco-Roman "Paganism" dawn of "civilization" saw a firm rejection of earlier views about the nature of life in the cosmos, and this included an aversion to Panspermia. Giordano Bruno's famous assertions on alien life that led to his death in 1600, and similar assertions have punctuated history. The modern interest in aliens expressed in science fiction can generally be interpreted as an implicit acknowledgement of alien life, and furthermore of our own intimate connection with it.

It is a curious fact that whilst science fiction movies portraying alien life become box office hits, the vast body of technical data on alien life such as been curated in this book are scarcely known to the public at large. This is not due to any weakness or inadequacy of the evidence itself, but due to the fact that prominent science journals that enjoy the widest media exposure deliberately act as censors of "unacceptable" science. In such a situation no discoveries get to be published except those that conform to the academic orthodoxy. We thus have a situation of total and absolute control of information in science and this is surely to be regretted.

The alternate viewpoint reviewed in this book that viral and bacterial genes are continually added to genomes of evolving life-forms leaves the role of neo-Darwinian evolution occupying a less important role as a kind of fine-tuning, rather than as the main driving force. Major evolutionary traits in the development of complex life are thus all externally derived, and evolution essentially driven from outside. If this is so the overall impression will be of a pre-programming in the higher levels of development in biology. The mechanism is that the relevant viral genes that were transported to Earth (neopanspermia, or even necropanspermia) had evolved over vast timescales and in innumerable locations and suddenly came to be expressed locally on the Earth. The evolution of the eye may be seen as one example of this type, and even some highly complex, and less definable manifestations of gene expression in our own immediate line of descent in hominid evolution bear the signs of "pre-programming" or pre-evolution as pointed out by the Japanese biologist, S. Ohno who, in 1970 wrote:

Did the genome of our cave-dwelling predecessor contain a set or set of genes which enabled modern man to compose music of infinite complexity and write novels with profound meaning? One is compelled to give an affirmative answer......It looks as though the early *Homo* was already provided with the intellectual potential which was in great excess of what was needed to cope with the environment of his time...

It can be speculated, consistent with the panspermia theory reviewed in this book, that the relevant genes for all such processes, including the musical and intellectual development in humans discussed by Ohno, evolved in a cosmic context and involved dispersal and exchange of viral genes across billions of light years. We can argue that Darwinian evolution occurred not on a single planet like Earth but over innumerable habitats in the grandest possible cosmic setting. We saw in earlier Chapters that evidence from the examination of meteoroids and other astronomical data supports this point of view. The huge numbers of habitable planets (over one hundred billion in our galaxy alone) that have recently been detected and the estimated mean distance of such planets being only a few light-years, imply that panspermia must be regarded as inevitable.

From studies of aquatic environments, it is estimated that about 10^{31} viruses are always in our midst, and the bulk of this grand total is likely to have fallen from the skies. Such viruses can have played three well-attested roles throughout our evolutionary history. They could cull us through the mechanism of disease (viral multiplication); they could protect us as in the role of placental syncytium, a virus-caused cellular membrane, that protects a fertilised egg (non-self) from being attacked by the immune system; or they could merge with our DNA (endogenisation), endowing it with evolutionary potential. They would also inevitably contribute to our microbiomes and thus have a controlling influence on our lives. If all such viruses — including those that contribute to our microbiomes — are of cosmic origin, it is not surprising that ingrained in every human being there is an instinctive feeling of a deep connection with some major aspect of the Universe.

Fig. 15.1. Gauguin's painting of 1897: "Where do we come from? What are we? Where are we going?".

Early man gave expression to this feeling by inventing pantheons of gods that mostly resided in the sky; and later religions inherited similar traditions that continue to the present day.

Artists through the ages also depicted a similar cosmic connection, as for instance in the famous painting in 1897 by Paul Guaguin with the title: "Where do we come from? What are we? Where are we going?" (Fig. 15.1) As clearly seen in the reproduction, in the reverential gaze upwards to the sky, tells the whole story.

As for the answers to Gauguin's questions, the first "Where do we come from?" is already answered. We (our DNA) came from space in the form of bacteria and viruses, cosmic viruses that evolved an almost infinite range of evolutionary outcomes over aeons of time. The answer to Gauguin's second question — what are we? — is thus also, at least partially, answered by virtue of developments in biology that have taken place after the dawn of the millennium. After the human genome was fully sequenced our cosmic ancestry was essentially laid bare. We have discovered as we have reviewed throughout this book that much of our genetic inheritance may be comprised of DNA actually delivered by viruses. At least 43% of the entire human genome may be traced directly to viruses and their closely related products. This estimate is based on the likely hypothesis that 34% of the human genome which constitute LINEs (Long Interspersed

Nuclear Elements) (21%), and SINEs (Short Interspersed Nuclear Elements) (13%), which are retroviral derived and controlled, and HERVs (Human Endogenous Retroviruses) and LTRs (Long Terminal Repeats) (9%).

We can now answer Gauguin's second question with confidence. We are essentially a complex of cosmic viruses; our evolution in every significant respect was directed by the ongoing incidence of cosmic viruses. In the fullness of time, long before the Sun exhausts its hydrogen fuel and becomes a red giant, some part of our augmented and reshuffled genomes will return to space to influence the evolution of life elsewhere in distant cosmic habitats. "Seeds" carrying our Earthly genetic heritage will eventually come to be sown on other planets.

Gauguin's third question "Where are we going to?" has the answer "We are going back into space — back into the cosmos where we came from."

Appendix

Text of Guest Editorial in International Journal of Astrobiology

International Journal of Astrobiology 1(2): 77–78 (2002) Printed in the United Kingdom
DOI: 10.1017/S1473550402001118 © 2002 Cambridge University Press

Editorial

The Beginnings of Astrobiology

With the present surge of interest in astrobiology and its emergence as a new scientific discipline in its own right, the role of a celebrated pioneer is all too often forgotten. There can be little doubt that the late Sir Fred Hoyle played a key part in relating astronomical phenomena to questions of life. One of his first contributions in this area was his introduction of the so-called anthropic principle to astronomy. By the late 1940's astronomers had worked out how the simplest chemical element Hydrogen could be converted into Helium in stars, thus providing the main energy source by which stars shine. The building of nuclei beyond Helium by stellar nuclear processes appeared difficult at the time because of instabilities in nuclei with atomic masses 5 and 8. Hoyle had the grand vision of making most if not all of the elements in the Periodic Table in stars. In the early 1950's Hoyle argued that by the very fact of our existence, the existence of life, the element Carbon had to be synthesised in quantity in stars. This could not happen, Hoyle concluded, unless the nucleus of Carbon possessed an energy level corresponding to a hitherto unknown excited state which he was able to calculate. This was necessary so that three Helium nuclei could combine first to form a Carbon nucleus in the excited state that subsequently decayed into the ground state. One of the major triumphs of Hoyle's Anthropic Principle was that his predicted excited state was subsequently discovered in the laboratory by Ward Whaling and Willy Fowler at Caltech. This discovery opened the door to a brand new discipline of Nuclear Astrophysics. In a seminal paper published in 1957, Hoyle together with Willy Fowler, Geoffrey and Margaret Burbidge showed that all the chemical elements needed for life C, N, O, P, Mg, Fe, S,... were made in stars. In a sense Hoyle's work in 1957 already provided the foundation stone for astrobiology. He showed that in essence we were made of stardust.

Fred Hoyle was amazingly prescient in recognising the importance of molecules in interstellar space even in the 1940's. Long before the discovery of the 21 cm line of neutral Hydrogen by Radio Astronomers Hoyle had argued for the widespread occurrence of Hydrogen molecules in the galaxy. Over two decades had to elapse before the molecule H_2 was discovered observationally and shown to be a major component of interstellar clouds.

In 1924 the Russian A.I. Oparin had promulgated the Primordial Soup Theory of the origin of life, and the same theory was independently proposed by the English Biologist J.B.S. Haldane in 1929. Hoyle admitted to being suspicious of Haldane's theory from the outset, as indeed he was of his politics! As early as 1955 in his classic book *Frontiers of Astronomy*, Hoyle discussed the merits of expanding the setting for the primordial soup to encompass the entire solar nebula, thus enhancing enormously the chances of life emerging from non-living material.

With the discoveries of carbon-based molecules in space, Hoyle and I began to consider even grander cosmic vistas for life. In the mid-1970's we argued that complex organic polymers would form on the surfaces of interstellar dust. We referred to polyformaldehyde, polysaccharides, bicyclic and polycyclic aromatics as prebiotic components of dust, identifying these classes of substances by infrared and ultraviolet spectral features. Although our own identifications in the 1970's were refuted at the time, similar identifications of organic and prebiotic polymers in interstellar dust are now generally accepted without dissent. By 1980 Hoyle and I made out a case for over 20% of interstellar carbon to be in the form of dust grains that mimicked the properties of freeze-dried bacteria. In view of the efficiency of conversion that was demanded by the astronomical observations we argued further that these organic dust particles most probably had a biological provenance. The idea was that comets condensing around primitive stellar-planetary systems inevitably mopped up a fraction of viable microbes from interstellar clouds, amplified them enormously in their warm watery interiors and returned most of it back into interstellar space from which new star systems can form. Some fraction of this biologically processed material would also and its way onto the surfaces of planets that could then be seeded with life. With the explorations of Comet Halley in 1986 the simple model of an inorganic dirty ice comet had to be replaced by one that included a significant organic component. Once again astronomical observations revealed that the cosmic dust from this comet was largely indistinguishable from bacterial material or their degradation products.

These ideas, referred to as panspermia, are still controversial of course, but there are definite signs of progress towards an acceptance, albeit in a limited form. It is now generally conceded that organics in interstellar clouds and in comets played a decisive role in life's origins on the Earth. Moreover, it is accepted that microorganisms are sturdy enough to withstand the rigors of space travel, if suitably coated or encased, so transfers of life between comets and planets in the solar system are more or less taken for granted.

Fred Hoyle's visions for astrobiology remained for the most part unrealised throughout his lifetime. Now they are slowly coming to be accepted and may even be drifting into mainstream science. His priority in this area is amply documented and beyond dispute. In a lecture delivered in Cardiff on 15 April 1980 entitled "The relation of biology to astronomy" Fred Hoyle concluded thus:

> Microbiology may be said to have had its beginnings in the nineteen forties. A new world of the most astonishing complexity began then to be revealed. In retrospect I find it remarkable that microbiologists did not at once recognise that the world into which they had penetrated had of necessity to be of cosmic order. I suspect that the cosmic quality of microbiology will seem as obvious to future generations as the Sun being the centre of our solar system seems obvious to the present generation.

Today Hoyle's prophesy may not be too far from being realised.

Chandra Wickramasinghe
Cardiff Centre for Astrobiology
Cardiff University

References

Relevant technical papers are reprinted in "Astronomical Origins of Life: Steps towards panspermia" ed. F. Hoyle and N.C. Wickramasinghe (Kluwer Academic Press, 2000).

Popular exposition: "Cosmic Dragons: Life and Death on our Planet" — (Souvenir Press, 2001).

Original invited Guest Editorial was first published in International Journal of Astrobiology, 1(2): 77–78 (2002)

Acknowledgements

In addition to the collaborators mentioned above, MW would like to thank Dr. Jim Gilmour and Professor Roger Anderson for useful discussions and support. We also thank Lakshmi Narayanan at WSPC for her patience in dealing with a difficult MS.

Biographies of Authors

Milton Wainwright, BSc, PhD, FRAS was born in 1950 in the mining village of Fitzwilliam in the West Riding of Yorkshire. He obtained his BSc and PhD from Nottingham University, and after a short period as a National Research Council of Canada Research Fellow became lecturer in Environmental Microbiology at the University of Sheffield. Here, he taught and researched for forty-two years in the Departments of Microbiology and Molecular Biology and Biotechnology. He is an Honorary Professor at the Universities of Cardiff and Buckingham, UK, the University of Ruhuna, Sri, Lanka, and the Slavic University of North Macedonia; he is also a Visiting Professor of King Saud University, Riyadh, and one of the few biologists to be made a Fellow of the Royal Astronomical Society. He has published widely on the history of science, particularly on the germ theory, the history of antibiotics (notably penicillin) and alternative accounts of the history of natural selection and evolution.

Nalin Chandra Wickramasinghe, MBE, BSc (Ceylon), MA, PhD, ScD (Cantab), Hon DSc (Sri Lanka, Ruhuna), Hon DLitt (Tokyo, Soka), FRAS, FRSA was born in 1939 in Sri Lanka. He commenced work in Cambridge on his PhD degree under the supervision of the late Sir Fred Hoyle, and published his first scientific paper in 1961. He was awarded a PhD degree in Mathematics in 1963 and was elected a Fellow of Jesus College Cambridge in the same year. In the following year he was appointed a Staff Member of the Institute of Astronomy at the University of Cambridge where he remained until 1973. He was formerly a Fellow of Jesus College Cambridge and Staff Member of the Institute of Astronomy, University of Cambridge; Formerly Professor and Head of the Department of Applied Mathematics and Astronomy, Cardiff University, UK; Director of the Buckingham Centre for Astrobiology, University of Buckingham, UK; Honorary Professor, University of Buckingham; Honorary Professor University of Ruhuna, Sri Lanka; Honorary Professor, Sir John Kotelawala Defence University of Sri Lanka; Adjunct Professor, National Institute of Fundamental Studies, Sri Lanka He has also held visiting Professorial appointment in the US, Canada and Japan and Sri Lanka over the past four decades. Professor Wickramasinghe has published over 350 papers in major scientific journals, some sixty in the journal *Nature*. Together with the late Sir Fred he pioneered the theory of cometary panspermia the evidence for which has become compelling over the past few years. Finally, he is also the author/co-author of over thirty-five books.

Further Reading

Cairns–Smith, A.G. (1982). *Genetic Takeover and The Mineral Origins of Life*, Cambridge University Press, Cambridge.

Fiore, M. (2022). *Prebiotic Chemistry and Life's Origins*. London, Royal Society of Chemistry.

Hoyle, F. and Wickramasinghe, N.C. (1978). *Lifecloud. The Origin of Life in the Galaxy*. London, Dent.

Hoyle, F. and Wickramasinghe, C. (1979). *Diseases from Space*. London, Dent.

Hoyle, F. and Wickramasinghe, N.C. (1981). *Space Travellers: The Bringers of Life*, Cardiff, University of Cardiff Press.

Hoyle, F. and Wickramasinghe, N.C. (1984). *From Grains to Bacteria*: Cardiff, University of Cardiff Press.

Hoyle, F. and Wickramasinghe, C. (1982). *Evolution from Space*. London, Dent.

Hoyle, F. and Wickramasinghe, N.C. (1985). *Living Comets*. Cardiff, University of Cardiff Press.

Hoyle, F., Wickramasinghe, N.C. and Watkins, J. (1986). *Viruses from Space*. Cardiff, University of Cardiff Press.

Hoyle, F. and Wickramasinghe, N.C. (1988). *Cosmic Life Force*. London, Dent.

Hoyle, F. and Wickramasinghe, N.C. (1990). *The Theory of Cosmic Grains*. Dordrecht, Kluwer.

Hoyle, F. and Wickramasinghe, N.C. (1993). *Our Place in the Cosmos*. London, Weidenfeld and Nicholson.

Hoyle, F. and Wickramasinghe, N.C. (2000). *Astronomical Origins of Life*. Dordrecht, Kluwer.

Luisi, P.L. (2010). *The Emergence of Life*. Cambridge, Cambridge University of Cambridge Press.

Ryan, F. (2009). *Virolution*. London, Harper Collins.

Wickramasinghe, C. (2015). *A Journey with Fred Hoyle*. Singapore, World Scientific Publishing.

Wickramasinghe, C. (1967). *Interstellar Grains*. London, Chapman and Hall.

Wickramasinghe, J., Wickramasinghe, C. and Napier, W. (2010). *Comets and the Origin of Life*. Singapore, World Scientific Publishing.

Wickramasinghe, C. (2001). *Cosmic Dragons: Life and Death on Our Planet*. London, Souvenir Press.

Wickramasinghe, J.T., Wickramasinghe, C., Napier, W.M. (2009). *Comets and the Origin of Life*. Singapore, World Scientific Publishing.

Wickramasinghe, C. (2015). *The Search for Our Cosmic Ancestry*, Singapore, World Scientific Publishing.

Wickramasinghe, C. (ed.) (2015). *Vindication of Cosmic Biology: A Tribute to Sir Fred Hoyle*, Singapore, World Scientific Publishing.

Walker, T. and Wickramasinghe, C. (2015). *The Big Bang and God*, USA Palgrave Macmillan.

Wickramasinghe, C. and Bauval, R. (2018). *Cosmic Womb*, USA, Inner Traditions.

Wickramasinghe, C., Wickramasinghe, K. and Tokoro, G. (2019). *Our Cosmic Ancestry in the Stars*, USA, Inner Traditions.

Wickramasinghe, C. (2015). *A Destiny of Cosmic Life — Chapters in the Life of an Astrobiologist*, Singapore, World Scientific Publishing.

Wickramasinghe, C. (2018). *Proofs that Life is Cosmic: Acceptance of a New Paradigm*, Singapore, World Scientific Publishing.

Mesler, B. and Cleaves, H.J. (2016). *A Brief History of Creation*, New York, Norton.

References to Technical Literature by Subject

(Note, C. and N.C. Wickramasinghe are the same author)

GENERAL PANSPERMIA

ALSHAMMARI, F. (2011). Evidence in support of the theory of archipanspermia. *Journal of Food, Agriculture and Environment*, 9, 1082–1084.

CALLAHAN, M.P., SMITH, K.E., CLEAVES, H.J., RUZICKA, J., STERN, J.C., GLAVIN, D.P., HOUSE, C.H. & DWORKIN, J.P. (2011). Carbonaceous meteorites contain a wide range of extraterrestrial nucleobases. *Proceedings of the National Academy of Sciences*, 108, 13995–13998.

CRICK, F.H. & ORGEL, L.E. (1973). Directed Panspermia. *Icarus*, 19, 341–346.

MARTINS, Z., BOTTA, O., FOGEL, M.L., SEPHTON, M.A., GLAVIN, D.P., WATSON, J.S., DWORKIN, J.P., SCHWARTZ, A.W. & EHRENFREUND, P. (2008). Extraterrestrial nucleobases in the Murchison meteorite. *Earth and Planetary Science Letters*, 270, 130–136.

NAPIER, W. (2004). A mechanism for interstellar panspermia. *Monthly Notices of the Royal Astronomical Society*, 348, 46–51.

NICHOLSON, W.L., MUNAKATA, N., HORNECK, G., MELOSH, H.J. & SETLOW, P. (2000). Resistance of *Bacillus* endospores to extreme terrestrial and extraterrestrial environments. *Microbiology and Molecular Biology Reviews*, 64, 548–572.

PAVLOV, A.K., KALININ, V.L., KONSTANTINOV, A.N., SHELEGEDIN, V.N. & PAVLOV, A.A. (2006). Was Earth ever infected by martian biota? Clues from radioresistant bacteria. *Astrobiology*, 6, 911–918.

RAULIN-CERCEAU, F., MAUREL, M.C. & SCHNEIDER, J. (1998.) From Panspermia to bioastronomy, the evolution of the hypothesis of universal life. *Origins of Life and Evolution of Biospheres*, 28, 597–612.

TAN, W. & VANLANDINGHAM, S. (1967). Electron microscopy of biological-like structures in the Orgueil carbonaceous meteorite. *Geophysical Journal of the Royal Astronomical Society*, 12, 237–237.

TIATIMMONS, D.E., FULTON, J.D. & MITCHELL, R.B. (1966). Micro-organisms of the upper atmosphere I. Instrumentation for isokinetic air sampling at altitude. *Applied microbiology*, 14, 229–231.

VALTONEN, M., NURMI, P., ZHENG, J.-Q., CUCINOTTA, F.A., WILSON, J.W., HORNECK, G., LINDEGREN, L., MELOSH, J., RICKMAN, H. & MILEIKOWSKY, C. (2009). Natural transfer of viable microbes in space from planets in extra-solar systems to a planet in our solar system and vice versa. *Astrophysical Journal*, 690, 210.

VREELAND, R.H., ROSENZWEIG, W.D. & POWERS, D.W. (2000). Isolation of a 250 million-year-old halotolerant bacterium from a primary salt crystal. *Nature*, 407, 897–900.

METEORITES

HOOVER, R.E., FRONTASYEVA. PAVLOV, S., ROZANOV, A. & WICKRAMASINGHE, N.C. (2021). ENNA and SEM investigaions of carbonaceous meteorites: Implications to the distribution of life and biospheres. *Academia Journal of Scientific Research*, 9, 96–104.

MCKAY, D.S., GIBSON JR, E.K., THOMAS-KEPRTA, K.L., VALI, H., ROMANEK, C.S., CLEMETT, S.J., CHILLIER, X.D., MAECHLING, C.R. & ZARE, R.N. (1996). Search for past life on Mars: possible relict biogenic activity in Martian meteorite ALH84001. *Science*, 273, 16.

MELOSH, H. (2003). Exchange of meteorites (and life?) between stellar systems. *Astrobiology*, 3, 207–215.

SEARS, D.W. & KRAL, T.A. (1998). Martian "microfossils" in lunar meteorites? *Meteorites and Planetary Science*, 33, 791–794.

STAPLIN, F.L. (1962). Microfossils from the Orgueil meteorite. *Micropaleontology*, 343–347.

THOMAS-KEPRTA, K.L., CLEMETT, S.J., BAZYLINSKI, D.A., KIRSCHVINK, J.L., MCKAY, D.S., WENTWORTH, S.J., VALI, H., GIBSON JR, E.K. & ROMANEK, C.S. (2002). Magnetofossils from ancient Mars: A robust biosignature in the Martian meteorite ALH84001. *Applied and Environmental Microbiology*, 68, 3663–3672.

WAINWRIGHT, M., ROSE, C., OMAIRI, T., BAKER, A., WICKRAMASINGHE, C. & ALSHAMMARI, F. (2014). A presumptive fossilized bacterial biofilm occurring in a commercially sourced Mars meteorite. *Astrobiol Outreach*, 2, 2332–2519.1000.

WAINWRIGHT, M., WICKRAMASINGHE, N.C. & TOKORA (2021). Polonnaruwa stones revisited — Evidence for non-terrestrial life. *Advances in Astrophysics*, 6, 19–25.

WALLACE, J. (2014). *Panspermia from Astronomy to Meteorites*. PhD Thesis, Cardiff University.

WICKRAMASINGH, N.C., WALLIS, J. & WALLIS, D.H. (2013). Panspermia: Evidence from astronomy to meteorites. *Modern Physics Letters*, 28, 1330009-1-13009-18.

WILLERSLEV, E., HANSEN, A.J., RØNN, R., BRAND, T.B., BARNES, I., WIUF, C., GILICHINSKY, D., MITCHELL, D. & COOPER, A. (2004). Long-term persistence of bacterial DNA. *Current Biology*, 14, R9–R10.

STRATOSPHERE

DELLA CORTA, V., RIETMEYER, F., ROTUNDI, A. & FERRARI, M. (2014). Intoducing a new stratopshere dust-collecting system with potemtial for upper atmosphere microbiology investigations. *Astrobiology*, 14, 694–705.

COLANGELI, L. (2016). Measurement of dust properties in different Solar System environments *Mem.S.A.It. Suppl.* 9, 161.

JUNGE, C.E. (1963). *Air Chemistry and Radioactivity*. New York, McGraw-Hill.

PAOLETTI, M.G. (ed.) (1999). *Invertebrate Biodiversity as Bio-indicators of Sustainable Landscapes* New York, Elsevier.

PLANE, J.M. (2012). Cosmic dust in the earth's atmosphere. *Chemical Society Reviews*, 41, 6507–6518.

ROSEN, J.M. (1969). Stratospheric dust and its relaitionship to the meteoric influx. *Space Science Review*, 9, 58–89.

SHIVAJI, S., CHATURVEDI, P., BEGUM, Z., PINDI, P.K., MANORAMA, R., PADMANABAN, D.A., SHOUCHE, Y.S., PAWAR, S., VAISHAMPAYAN, P. & DUTT, C. (2009). *Janibacter hoylei* sp. nov., *Bacillus isronensis* sp. nov. and *Bacillus aryabhattai* sp. nov., isolated from cryotubes used for collecting air from the upper atmosphere. *International Journal of Systematic and Evolutionary Microbiology*, 59, 2977–2986.

SMITH, D.J. (2013). Microbes in the upper atmosphere and unique opportunities for astrobiology research. *Astrobiology*, 13, 981–990.

SMITH, D.J., GRIFFIN, D.W. & SCHUERGER, A.C. (2010). Stratospheric microbiology at 20 km over the Pacific Ocean. *Aerobiologia*, 26, 35–46.

WAINWRIGHT, M. (2003). A microbiologist looks at panspermia. *Astrophysics and Space Science*, 285, 563–570.

WAINWRIGHT, M. (2008). The high cold biosphere — Microscope studies on the microbiology of the stratosphere. *In Focus-Proceedings of the Royal Microscopical Society*, 33–41.

WAINWRIGHT, M., ALHARBI, S. & WICKRAMASINGHE, N.C. (2006). How do microorganisms reach the stratosphere? *International Journal of Astrobiology*, 5, 13.

WAINWRIGHT, M., ROSE, C., BAKER, A., WICKRAMASINGHE, N. & OMAIRI, T. (2015). Biological entities isolated from two stratosphere launches-continued evidence for a space origin. *Astrobiology Outreach*, 3, 2332–2519.1000.

WAINWRIGHT, M., WICKRAMASINGHE, N., HARRIS, M. & OMAIRI, T. (2015). Masses staining positive for DNA-isolated from the stratosphere at a height of 41 km. *Astrobiology Outreach*, 3, 2332–2519.1000.

WAINWRIGHT, M., WICKRAMASINGHE, N., NARLIKAR, J., RAJARATNAM, P. & PERKINS, J. (2004). Confirmation of the presence of viable but non-cultureable bacteria in the stratosphere. *International Journal of Astrobiology*, 3, 13–15.

WAINWRIGHT, M., WICKRAMASINGHE, N.C., NARLIKAR, J. & RAJARATNAM, P. (2003). Microorganisms cultured from stratospheric air samples obtained at 41 km. *Federation of European Microbiological Societies Letters*, 218, 161–165.

WAINWRIGHT, M. (2015). Biological entoties and DNA-containig k masses isolated form th strtosphere-evidence for anon-terrestrial origin. *Astronomy Rviews*, 1, 25–40.

WAINWRIGHT, M., WICKRAMASINGHE, N.C. & TOKORO, G. (2021). Neopanspermia-evidence that life continually arrives at Earth from space. *Advances in Astrophysics*, 6, 6–18.

YAMAGAMI, T., SAITO, Y., MATSUZAKA, Y., NAMIKI, M., TORIUMI, M., YOKOTA, R., HIROSAWA, H. & MATSUSHIMA, K. (2004). Development of the highest altitude balloon. *Advances in Space Research*, 33, 1653–1659.

YANG, Y., ITAHASHI, S., YOKOBORI, S.-I. & YAMAGISHI, A. (2008). UV-resistant bacteria isolated from upper troposphere and lower stratosphere. *Biological Sciences in Space*, 22, 7.

YANG, Y., YOKOBORI, S.-I., KAWAGUCHI, J., YAMAGAMI, T., IIJIMA, I., IZUTSU, N., FUKE, H., SAITOH, Y., MATSUZAKA, S. & NAMIKI, M. (2008). Investigation of cultivable microorganisms in the stratosphere collected by using a balloon in 2005. *JAXA Research and Development Report*, 35–42.

YANG, Y., YOKOBORI, S.-I. & YAMAGISHI, A. (2009). Assessing panspermia hypothesis by microorganisms collected from the high altitude atmosphere. *Biological Sciences in Space*, 23, 151–163.

Index

CPSIA information can be obtained
at www.ICGtesting.com
Printed in the USA
BVHW090733081222
653663BV00004B/7